PRACTICE MAKES PERFECT

eve 3

practice
sheets
& quizzes

MASTERBOOKS®
CURRICULUM

Author: Angela O'Dell

Master Books Creative Team:

Editor: Craig Froman

Design: Terry White

Cover Design:
Diana Bogardus

Copy Editors:
Judy Lewis
Willow Meek

Curriculum Review:
Kristen Pratt
Laura Welch
Diana Bogardus

First printing: November 2020

Master Books® is a division of the New Leaf Publishing Group, Inc.

ISBN: 978-1-68344-218-9
ISBN: 978-1-61458-765-1 (digital)

Images are from gettyimage.com, istock.com, and shutterstock.com.

Printed in the United States of America

Please visit our website for other great titles:

www.masterbooks.com

Author Bio:

As a homeschooling mom and author, **Angela O'Dell** embraces many aspects of the Charlotte Mason method yet knows that modern children need an education that fits the needs of this generation. Based upon her foundational belief in a living God for a living education, she has worked to bring a curriculum that will reach deep into the heart of home-educated children and their families. She has written over 20 books, including her history series and her math series. Angela's goal is to bring materials that teach and train hearts and minds to find the answers for our generation in the never-changing truth of God and His Word.

Scope and Sequence

Welcome to *Practice Makes Perfect Level 3*

Please carefully read through the following sections on how and when to use this optional *Math Lessons for a Living Education* supplemental product. It is necessary to have the main student book in order to complete these pages.

How to Implement *Practice Makes Perfect*

- After your student finishes with their lesson activity in their Math Lessons curriculum workbook, you, the parent, may decide to have them complete a little more practice.

- Please do not feel like you need to use every single activity page. Instead, choose activity pages based on the individual need of your student. If they need more practice or they would like to do more activity sheets, simply give them the page which meets their need.

- There are four quarterly quizzes included in each level of *Practice Makes Perfect*. Please remember, these are not mandatory. The oral narrations and the interactive nature of the Math Lessons for a Living Education curriculum is plenty for many families.

The Purpose and Goals of *Practice Makes Perfect*

- These extra practice pages are a resource for when a little extra practice is needed or wanted, and to give the families using the *Math Lessons for a Living Education* curriculum series helpful support in the form of four quarterly quizzes, which they can keep for their written records when such records are required by their state's educational laws.

Goals, Tips, and Focus for Review Lessons 31-36:

Lessons 31-36 are focused review lessons for the major concepts taught in this level of *Math Lessons for a Living Education*. Because these lessons are already focused reviews, there are no extra review pages in this *Practice Makes Perfect*. The goal for these lessons is for you, the parent, to be able to ensure your student has a good mastery of the concepts. To determine mastery, ask yourself these questions:

- Does my child show mastery through application? For example, can my child apply this concept in unrehearsed situations (not in their math book) to which I purposefully expose them?

- Does my child show mastery through real world connections? Do they purposefully and correctly use their math knowledge in real life?

How to make the most of the reviews:

As your child works through each of these review lessons, take the time to watch them interact with the concepts. Watch carefully how they interact with any manipulatives, the confidence they use when presenting any show-and-tell projects, and their ability to orally narrate their understanding of any and all of the concepts reviewed in these lessons. After they are finished with these review lessons, you have the option of having them complete the Quarter 4 Quiz.

Supply List

The following supplies are needed for completing these activities: crayons, colored pencils, bottle caps, and scissors.

Worksheet Section

Let's get started!

Name_____

Review of Place Value, Odds and Evens, Counting by 2s, 5s, 10s

Place Value Slider. Instructions for assembly are on the back.

0	0	0	0
1	1	1	1
2	2	2	2
3	3	3	3
4	4	4	4
5	5	5	5
6	6	6	6
7	7	7	7
8	8	8	8
9	9	9	9

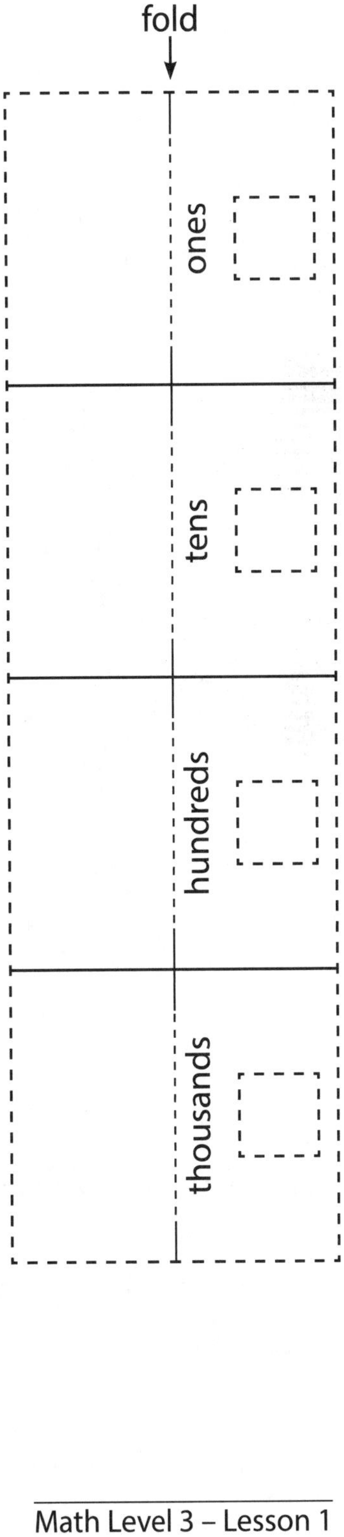

fold

ones

tens

hundreds

thousands

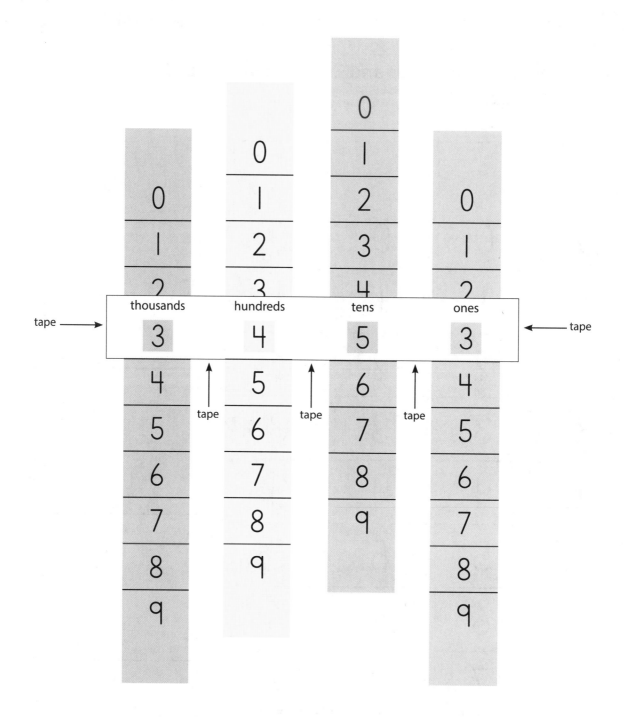

Instruction for assembly. Cut out along the dotted lines. Your teacher needs to cut out the white piece. Fold the white piece and tape the short edges (use extra tape where needed). Lay it flat to use. Practice creating numbers. **You will need this in later lessons.**

Name_____

Evens or Odds? Circle whether the number is odd or even. Even numbers always end in 0, 2, 4, 6, 8. Odd numbers always end in 1, 3, 5, 7, 9.

32	**490**	**33**
Odd Even	Odd Even	Odd Even
175	**82**	**919**
Odd Even	Odd Even	Odd Even
642	**45**	**777**
Odd Even	Odd Even	Odd Even

Name_____

Dot-to-Dot. Starting with the number 1, connect the dots up to number 56. Then color your picture.

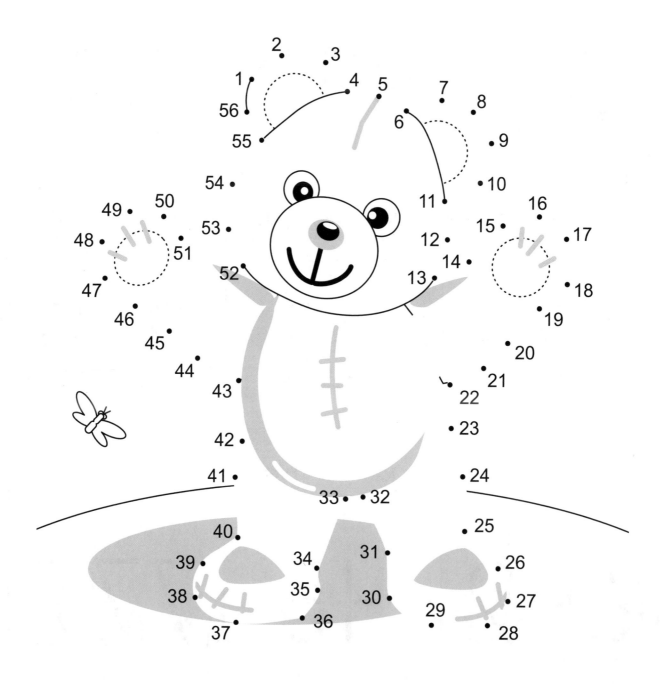

Name_____

Copywork of Numbers

100   101   102   103   104   105   106

107   108   109   110   111   112   113

114   115   116   117   118   119   120

121   122   123   124   125   126   127

128   129   130   131   132   133   134

135   136   137   138   139   140   141

142   143   144   145   146   147   148

149   150

Name_____

Copywork of Numbers

151   152   153   154   155   156   157

158   159   160   161   162   163   164   165

166   167   168   169   170   171   172   173

174   175   176   177   178   179   180

181   182   183   184   185   186   187

188   189   190   191   192   193

194   195   196   197   198   199   200

Name_____

Review of Money, Clocks, Perimeter, Addition/Subtraction Facts

Which clock is it? Circle the correct clock.

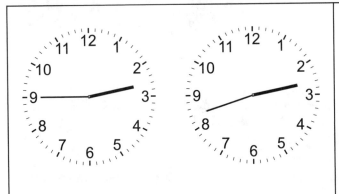

2:42

1:51

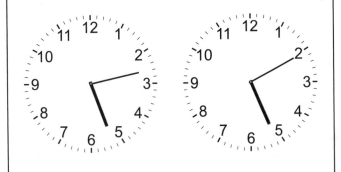

5:13

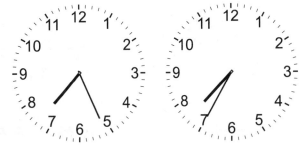

7:26

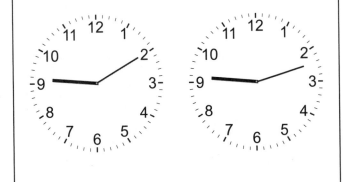

9:12

12:21

Welcome to the Amusement Park!

Figure out your budget for the carnival rides below. The first one is done for you.

$2.00 $3.00 $2.50 $1.50 $2.25

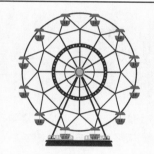 +  =
$2.00
+ $2.50
$4.50

 + 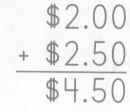 =

 +  =

Name_____

Review of Addition, Including Carrying, Tally Marks

Missing dots. Draw in the missing dots to make the addition equation true. The first one is done for you.

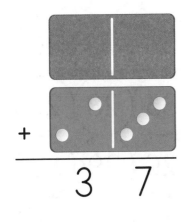

+

5   3

+

4   5

+

6   7

+

3   7

+

4   7

+

7   6

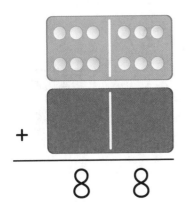

+

8   8

+

7   9

+

6   8

Name_____

Skip Counting by 5

Add 5 to the last blue bubble and color it blue to help the turtle find his way to the starfish at the end of the maze. Examples: $5 + 5 = 10$, $10 + 5 = 15$

2	1	6	8	3	10
➡	5	10	9	7	9

13	11	8	7	8	4	15	20	12	14
17	12	16	14	11	9	11	25	9	11
24	16	13	18	13	14	19	30	22	26
27	21	19	55	50	45	40	35	28	22
22	23	30	60	39	21	23	26	29	32
28	25	27	65	20	23	27	30	33	36
39	37	34	70	75	80	85	➡		
42	40	38	33	25	37	40	42		

Name_____

Treasure Chest Tally

Count the tally marks in each treasure chest and write the number in the box.

Name_____

Do the Math Challenge

Do the addition and subtraction to fill in this chart. The first one is done for you.

	1 More	1 Less	10 More	10 Less
12	13	11	22	2
18				
15				
20				
16				
40				

Review of Subtraction, Including Borrowing Concepts

Subtraction Race Game

To play the subtraction race game, you will need: two players (student plus 1), two pencils, scrap paper for both players, the included game pieces, a die (one dice), a timer, and a calculator.

Steps:

Player 1 rolls the die and moves their game piece that many spaces.

If they land on a problem, they set the timer for 1.5 minutes and try solve the problem (the other player checks it on a calculator). If they get it correct, they get to go another turn. Limit three turns in a row for each player. If at any point in the game, they land in a mud puddle, they have to go back to start.

Player 2 does the same thing.

To win, reach the finish line first.

Game board and game pieces are on the back.

Laminate gameboard so it can be used over and over.

START

700
− 234

612
− 589

465
− 399

930
− 841

577
− 259

MUD

300
− 197

800
− 337

901
− 658

856
− 792

200
− 136

437
− 344

291
− 263

MUD

931
− 518

265
− 178

601
− 158

998
− 769

MUD

FINISH

game pieces

Name_____

More Time-Telling Practice

Start

Finish

How long did it take Charlie and Hairo to clean up their room?

_____ minutes

Start

Finish

How much time did Charlotte spend reading?

_____ minutes

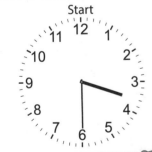
Start

Finish

How much time did Mom take reviewing phonics with Natty?

_____ minutes

Start

Finish

How long did it take for Dad to get his hair cut?

_____ minutes

Name_____

Earlier and Later

Looking at the clock in the middle, draw and write the correct times showing half an hour earlier and half an hour hour later.

½ hour earlier	present time	½ hour later

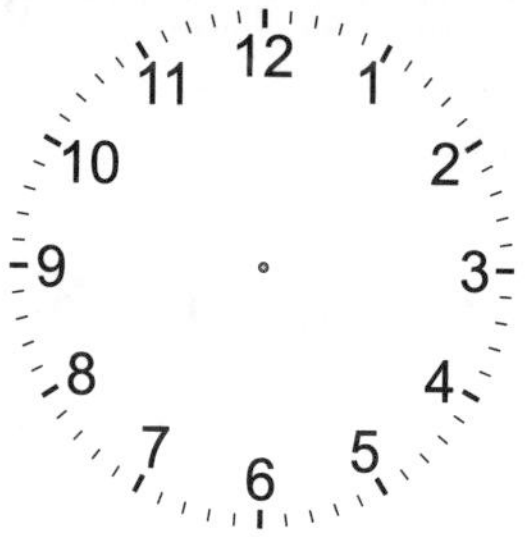

1:00

6:00

10:00

Review of Measurement, Fractions, Thermometers, Graphs

Measure the Silverware!

Measure each item with the ruler provided below it. The first one is done for you.

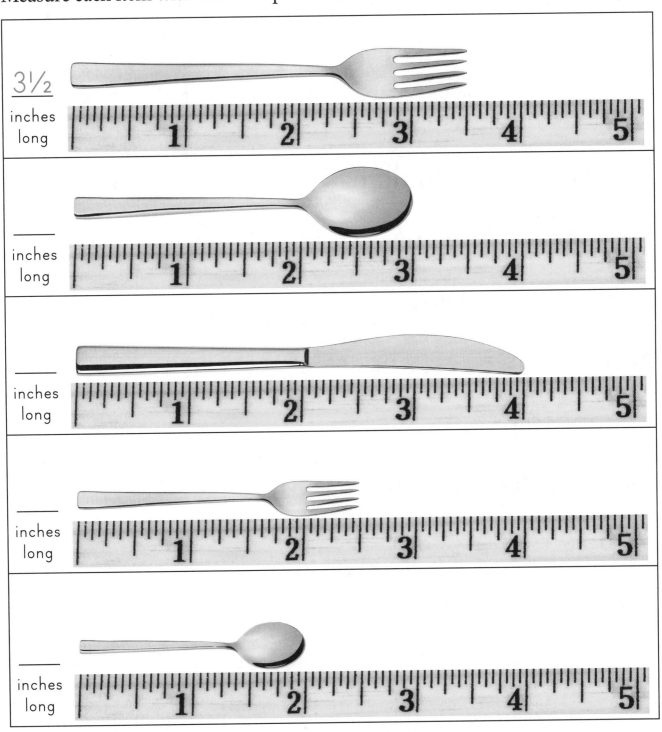

$3\frac{1}{2}$
inches long

_____
inches long

_____
inches long

_____
inches long

_____
inches long

Name_____

Practice time!

Fill in the blank with all you remember.

16 ounces = _____ pound

1 foot = _____ inches

3 feet = _____ yard

4 quarts = _____ gallon

1 quart = _____ pints

2 cups = _____ pint

60 minutes = _____ hour

12 months = _____ year

60 seconds = _____ minute

24 hours = _____ day

1 week = _____ days

Name_____

Count the Coins

Charlie has been practicing counting coins. Circle "thumbs up" if he gave the correct answer. Circle "thumbs down" if he gave the incorrect answer.

1¢   5¢   10¢   25¢

26¢   62¢

69¢   96¢

$1.23 $1.33

Name_____

Measuring with Cups, Pints, and Quarts

Look at the measurement chart to the right and then circle the correct amount below.

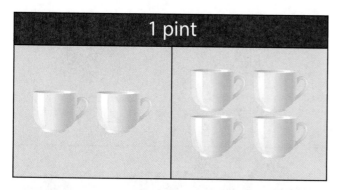

1 pint

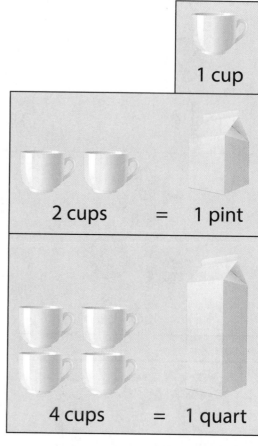

1 cup

2 cups = 1 pint

4 cups = 1 quart

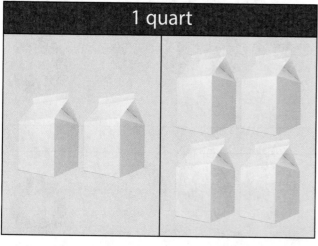

1 quart

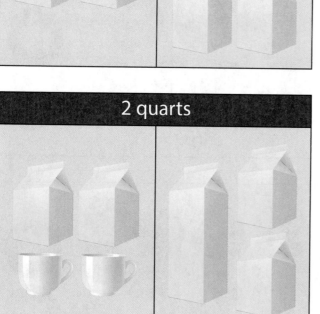

2 quarts

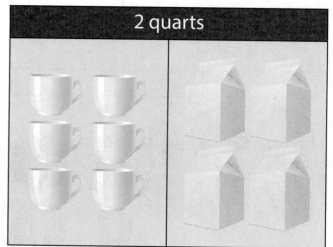

2 quarts

Name_____

Review of Word Problems

Since the entire lesson in your math book is centered around reviewing how to solve word problems, you may take some time each day and practice some other skills.

Ordering Numbers

Place the numbers in the correct order from smallest number to largest number.

332,	350,	312,	300
___,	___,	___,	___

643,	421,	390,	199
___,	___,	___,	___

454,	450,	452,	459
___,	___,	___,	___

900,	867,	712,	910
___,	___,	___,	___

231,	132,	321,	213
___,	___,	___,	___

745,	457,	547,	754
___,	___,	___,	___

101,	110,	220,	201
___,	___,	___,	___

874,	478,	847,	487
___,	___,	___,	___

Name_____

Yesterday, Today, and Tomorrow. Write down the day before and the day after for each day in the "Today" column.

Yesterday	Today	Tomorrow
	Thursday	
	Tuesday	
	Saturday	
	Sunday	
	Friday	
	Wednesday	
	Monday	

Addition and Subtraction Families Practice and Activity

Cut out the fact families circles, fold forward on the small dotted lines, and glue to a separate piece of notebook paper. See an example on the back.

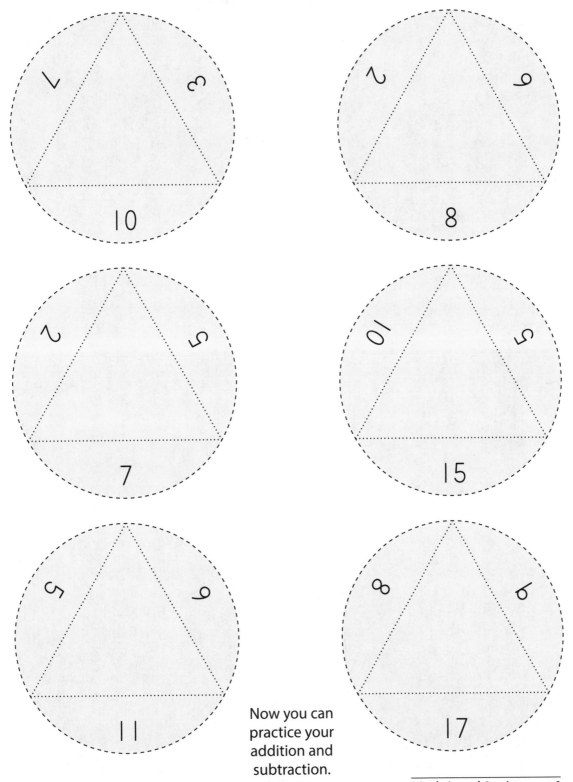

Now you can practice your addition and subtraction.

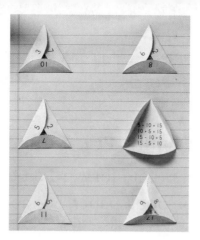

$2 + 6 = 8$
$6 + 2 = 8$
$8 - 2 = 6$
$8 - 6 = 2$

$3 + 7 = 10$
$7 + 3 = 10$
$10 - 3 = 7$
$10 - 7 = 3$

$5 + 10 = 15$
$10 + 5 = 15$
$15 - 10 = 5$
$15 - 5 = 10$

$2 + 5 = 7$
$5 + 2 = 7$
$7 - 2 = 5$
$7 - 5 = 2$

$8 + 9 = 17$
$9 + 8 = 17$
$17 - 8 = 9$
$17 - 9 = 8$

$5 + 6 = 11$
$6 + 5 = 11$
$11 - 5 = 6$
$11 - 6 = 5$

Name_____

Mutt Math: Subtraction

Today, let's practice subtraction facts. Subtract the numbers around the middle of the wheel from the number in the center. Write the answer on the outside ring.

Name_____

Calendar Practice

For each question below, draw a line to the correct month.

Third month of the year   January

Month after March   February

Month between June and August   March

Month before September   April

The last month of the year   May

Two months before March   June

Second month of the year   July

Fifth month of the year   August

Ninth month of the year   September

The month before July   October

The month after September   November

The month before the last month of the year   December

Name_____

How long and how much?

How long did it take Charlie and Hairo to peel and cut the potatoes for the supper cooks at the children's home?

_____ minutes

How long did it take the children to take showers, brush their teeth, and be ready for their bedtime story?

_____ minutes

How long did it take for the children to finish their penmanship practice pages?

_____ minutes

How long did it take the supper dishes crew to clear the tables after dinner at the children's home?

_____ minutes

Odds and Evens

Circle whether the number is odd or even.

26	**901**	**164**
Even Odd	Even Odd	Even Odd
75	**99**	**82**
Even Odd	Even Odd	Even Odd
415	**623**	**508**
Even Odd	Even Odd	Even Odd
317	**300**	**222**
Even Odd	Even Odd	Even Odd

Introducing Rounding to the 10s and 100s

Follow these steps to understand the idea of rounding.

Step 1: Write or think which tens the number is in between.

Step 2: Look at the digit in the ones' place. Is it greater than or less than 5?

Step 3: If the digit in the ones' place is 5 or greater than 5, round to the larger ten.

Step 4: If the digit in the ones' place is less than 5, round to the smaller ten.

Show your teacher how to round these numbers to the nearest 10. Circle the correct answer.

42   40 or 50

91   90 or 100

55   50 or 60

28   20 or 30

33   30 or 40

97   90 or 100

Name_____

Missing Addition Number

Write the missing addition number in the box. The first one is done for you.

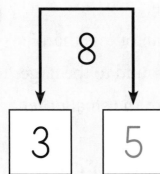

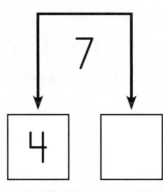

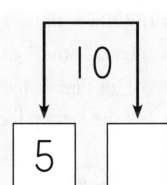

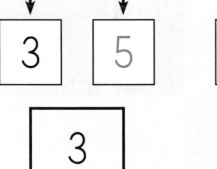

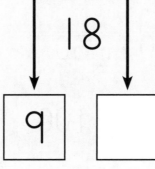

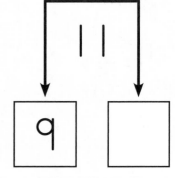

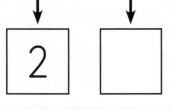

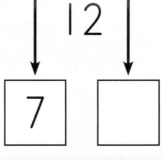

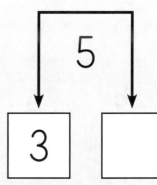

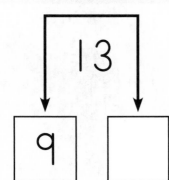

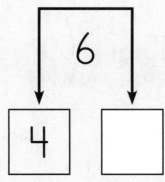

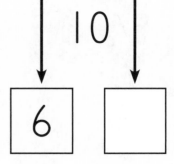

Name_____

Missing Addition Number

Write the missing addition number in the box.

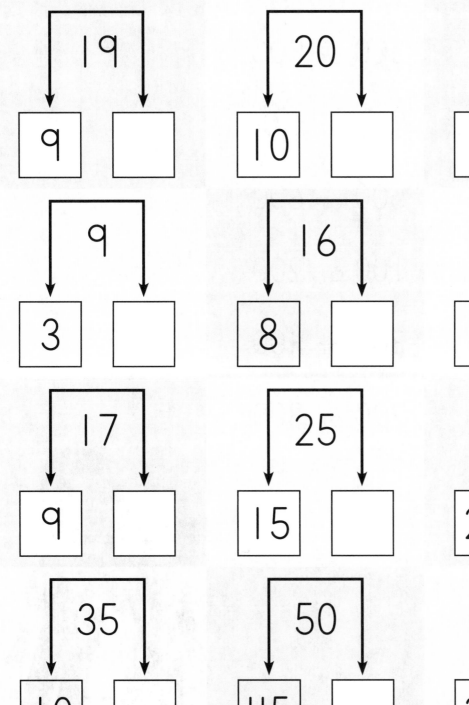
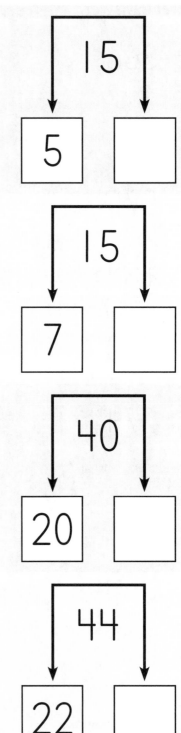

Rounding to the nearest 100

Show your teacher how to round these numbers to the nearest 100. Circle the correct answer.

366	300 or 400
931	900 or 1,000
575	500 or 600
183	100 or 200
338	300 or 400
274	200 or 300

Rounding and Mental Math Practice

Try to solve these in your brain.

Round these to the nearest 10. Tell your teacher the answers.

13 _____   46 _____

81 _____   97 _____

Round these to the nearest 100. Tell your teacher the answers.

846 _____   912 _____

277 _____   538 _____

High and Low

Write the numbers from least to greatest.

4,217   4,172   4,127   4,271

_____   _____   _____   _____

least   between   between   greatest

More or Less? – Number Sense

Do the addition and subtraction to fill in this chart. The first one is done for you.

	10 More	10 Less
65	75	55
12		
88		
52		
71		
47		
34		

	10 More	10 Less
29		
19		
63		
13		
85		
79		
90		

Name_____

Review of All New Concepts

Solve these addition problems. Show your work if you need to carry over.

$$
\begin{array}{r}
678 \\
+\ 203 \\
\hline
\end{array}
\qquad
\begin{array}{r}
780 \\
+\ 567 \\
\hline
\end{array}
\qquad
\begin{array}{r}
576 \\
+\ 408 \\
\hline
\end{array}
$$

$$
\begin{array}{r}
457 \\
232 \\
+\ 271 \\
\hline
\end{array}
\qquad
\begin{array}{r}
431 \\
523 \\
+\ 683 \\
\hline
\end{array}
\qquad
\begin{array}{r}
356 \\
439 \\
+\ 226 \\
\hline
\end{array}
$$

Create some of your own!

$$
\begin{array}{r}
+ \\
\hline
\end{array}
\qquad
\begin{array}{r}
+ \\
\hline
\end{array}
\qquad
\begin{array}{r}
+ \\
\hline
\end{array}
$$

$$
\begin{array}{r}
+ \\
\hline
\end{array}
\qquad
\begin{array}{r}
+ \\
\hline
\end{array}
\qquad
\begin{array}{r}
+ \\
\hline
\end{array}
$$

Solve these subtraction problems.

$$\begin{array}{r} 67 \\ -\ 58 \\ \hline \end{array}$$
$$\begin{array}{r} 381 \\ -272 \\ \hline \end{array}$$
$$\begin{array}{r} 805 \\ -123 \\ \hline \end{array}$$

$$\begin{array}{r} 208 \\ -124 \\ \hline \end{array}$$
$$\begin{array}{r} 834 \\ -\ 54 \\ \hline \end{array}$$
$$\begin{array}{r} 673 \\ -\ 456 \\ \hline \end{array}$$

Create some of your own!

$$\begin{array}{r} - \\ \hline \end{array}$$
$$\begin{array}{r} - \\ \hline \end{array}$$
$$\begin{array}{r} - \\ \hline \end{array}$$

$$\begin{array}{r} - \\ \hline \end{array}$$
$$\begin{array}{r} - \\ \hline \end{array}$$
$$\begin{array}{r} - \\ \hline \end{array}$$

Name________________________

Figure it out!

Solve and round the answer to the nearest 100.

633 + 258 Rounds to ☐	387 + 435 Rounds to ☐
589 - 143 Rounds to ☐	348 - 139 Rounds to ☐

Rounding and In Between

Write the 10s the number is between, then write the ten the number is closer to.

46 is between _____ and _____ and closer to _____

74 is between _____ and _____ and closer to _____

83 is between _____ and _____ and closer to _____

Write the 100s the number is between, then write the hundred the number is closer to.

432 is between _____ and _____ and closer to _____

567 is between _____ and _____ and closer to _____

258 is between _____ and _____ and closer to _____

Name_____

Finding Fun Answers!

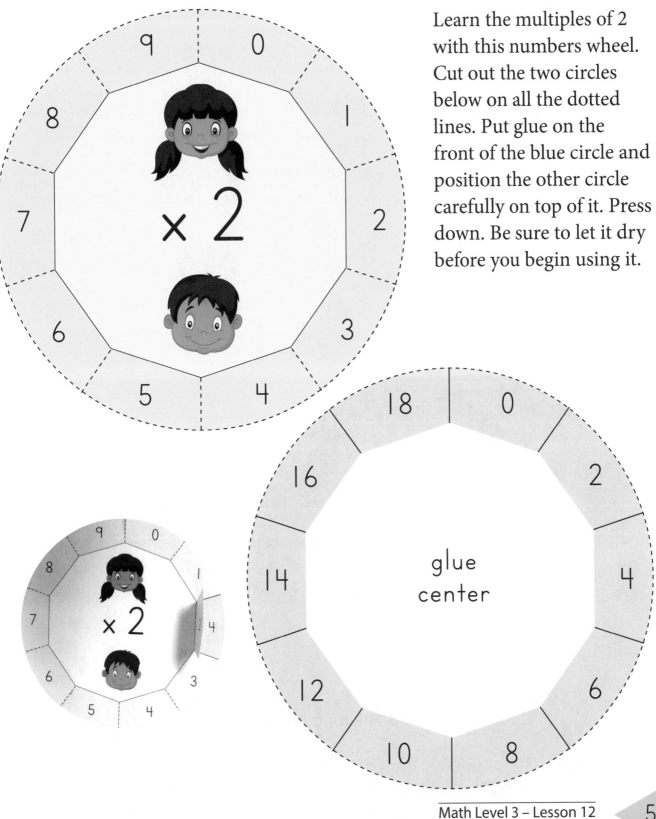

Learn the multiples of 2 with this numbers wheel. Cut out the two circles below on all the dotted lines. Put glue on the front of the blue circle and position the other circle carefully on top of it. Press down. Be sure to let it dry before you begin using it.

Blank for cutting.

Name_____

Introducing Division of 1, 2, and 5

Fact Family Practice. Cut out and color the houses provided. Slip them into a page protector or laminate them. Use them to practice at least four of the fact families that you learned in your math lesson. (See example on the back.)

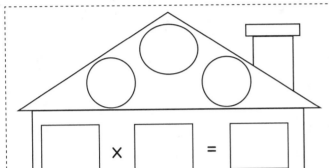

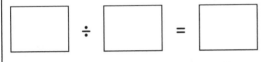

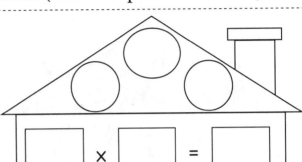

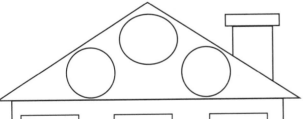

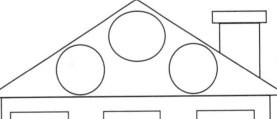

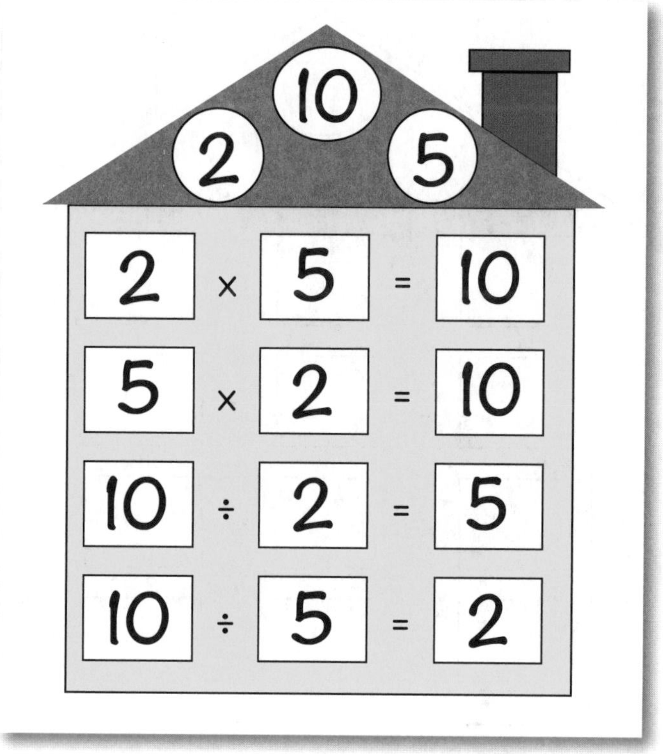

Fact family house with roof circles: 2, 10, 5

$2 \times 5 = 10$

$5 \times 2 = 10$

$10 \div 2 = 5$

$10 \div 5 = 2$

Name_____

Mutt Math: Multiplication

Today, let's practice multiplication facts. Multiply the numbers on the outside of the wheel by the number in the center. Put your answers in the outer circle.

Mutt Math: Division

Today, let's practice division facts. Divide the numbers on the outside of the wheel by the number in the center. Put your answers in the outer circle.

Name_____

Introducing the Area of Rectangles and Squares

Area File Folder Game. Laminate this page so you can use it again.

Each play toss two dice. Play with a friend. Highest number goes first.

On your turn:

- Toss the dice. Use one number for length and the other for width of the shape.

- Draw the shape and multiply its length and width to find its area.

Take turns until there is no room left on one or both of the game boards.

To win: whoever has more shapes on their game board wins.

Be sure to save these for Lesson 18.

(See examples on the back.)

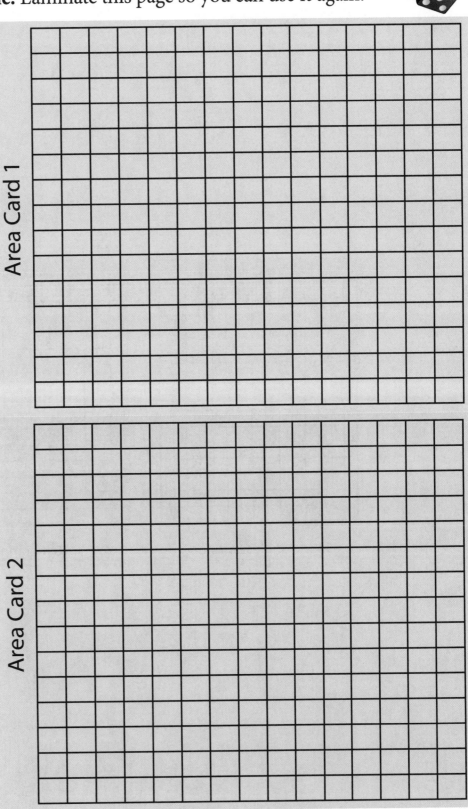

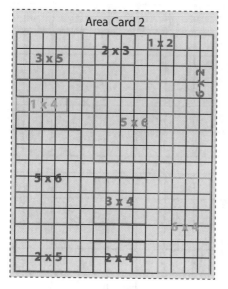

11 rectangles (winner)

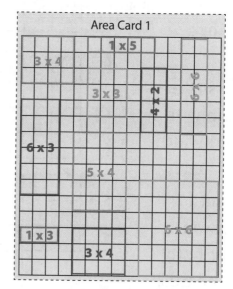

10 rectangles

Name_____

Math Crossword. Fill in the missing squares with the correct number.

2	×		=	10

+
5
=

10
+
10
=

10	÷	2	=	

| | × | 2 | = | |

×
5
=

×
3
=

3	×		=	15

2	×		=	20

20
÷
=
5

5	×		=	15

15
−
5
=

Multiplication and Division. Solve these multiplication problems. Underneath each one, write a matching division problem.

2 × 3 =   2 × 5 =   2 × 7 =

Mental Math Mania!

Solve each mental math number sentence. Work from left to right. Each time you get one right, do two jumping jacks. Tell your teacher the correct answer.

$10 \times 2 - 5 + 10 - 4 =$

$25 \div 5 + 10 + 5 + 10 \div 3 =$

$18 - 9 \times 10 + 5 + 3 + 2 =$

Area and Perimeter

Figure out the area of a square that has 5-foot sides.

Now, decide the perimeter of a rectangle that has a 10-yard side and an 8-yard side.

Name_____

Introducing Multiplying and Dividing by 3

Cut out the 3x study-buddy and use it to learn to multiply by 3. (See example on the back.)

	0×3
	1×3
	2×3
	3×3
	4×3
	5×3
	6×3
	7×3
	8×3
	9×3
	10×3

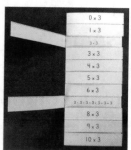

3

3 + 3

3 + 3 + 3

3 + 3 + 3 + 3

3 + 3 + 3 + 3 + 3

3 + 3 + 3 + 3 + 3 + 3

3 + 3 + 3 + 3 + 3 + 3 + 3

3 + 3 + 3 + 3 + 3 + 3 + 3 + 3

3 + 3 + 3 + 3 + 3 + 3 + 3 + 3 + 3

3 + 3 + 3 + 3 + 3 + 3 + 3 + 3 + 3 + 3

Rounding. Round these to the nearest 10.

32 _____   76 _____

91 _____   14 _____

29 _____   82 _____

Round these to the nearest 100.

423 _____   755 _____

910 _____   132 _____

377 _____   821 _____

How Long. Solve this story problem.

Hairo volunteered to sweep three of the hallways in the children's home. He started his chore at 1:20 in the afternoon and finished at 3:45. Draw hands on the first clock to show the time when he started. Draw hands on the second clock to show when he finished. How long did it take him to sweep the 3 hallways?

Counting

Use your money manipulatives to count out these amounts.

☐ $4.56

☐ $ .94

☐ $ .83

☐ $3.99

Challenge:

If someone gave you a $5 bill, and asked you to give them back $2.50, how would you figure out how to do this? Write the equation and solve it.

Introducing Multiplying and Dividing by 4

Finding Fun Answers!

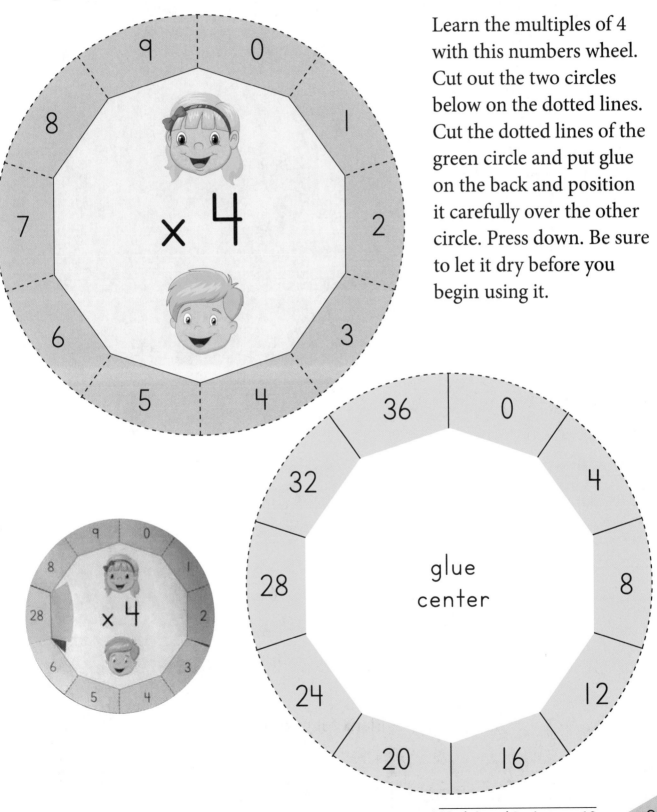

Learn the multiples of 4 with this numbers wheel. Cut out the two circles below on the dotted lines. Cut the dotted lines of the green circle and put glue on the back and position it carefully over the other circle. Press down. Be sure to let it dry before you begin using it.

Blank for cutting.

Name_____

Perimeter and Area Practice

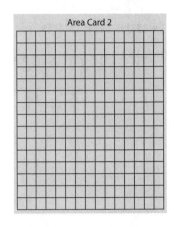

Area Card 2

Option 1:

Using your Area File Folder Game (which you created in Lesson 15) practice the concept of finding the area of rectangles and squares.

Option 2:

Practice the concept of finding the perimeter of squares using multiplication.

Here's how:

1. Using your feet, measure off 3 squares of different sizes.

2. Fill in the blanks below with your information and find the perimeter of each square. The first one is done for you.

Example Square:

Each side is ___3___ of your feet long. ___3___ x 4 = ___12___ of your feet
[This number will be the perimeter of your square.]

Square 1:

Each side is _____ of your feet long. _____ x 4 = _____ of your feet
[This number will be the perimeter of your square.]

Square 2:

Each side is _____ of your feet long. _____ x 4 = _____ of your feet
[This number will be the perimeter of your square.]

Square 3:

Each side is _____ of your feet long. _____ x 4 = _____ of your feet
[This number will be the perimeter of your square.]

Name the Sign

Write the missing signs.

$36 \_\_ 9 = 4$ $10 \_\_ 4 = 40$ $16 \_\_ 4 = 4$

$10 \_\_ 4 = 6$ $8 \_\_ 2 = 4$ $4 \_\_ 6 = 10$

Fill in the Blanks

Multiplying and dividing by 4.

$4 \times 1 = $ _____ $12 \div 4 = $ _____

$4 \times 3 = $ _____ $8 \div 4 = $ _____

$4 \times 6 = $ _____ $16 \div 4 = $ _____

$4 \times 9 = $ _____ $4 \div 4 = $ _____

$4 \times 4 = $ _____ $20 \div 4 = $ _____

Name_____

Introducing Multiplying and Dividing by 6 and 7

Learn the multiples of 6 with this numbers wheel. Cut out the two circles below on the dotted lines. Cut the dotted lines of the blue circle and put glue on the back and position it carefully over the other circle. Press down. Be sure to let it dry before you begin using it.

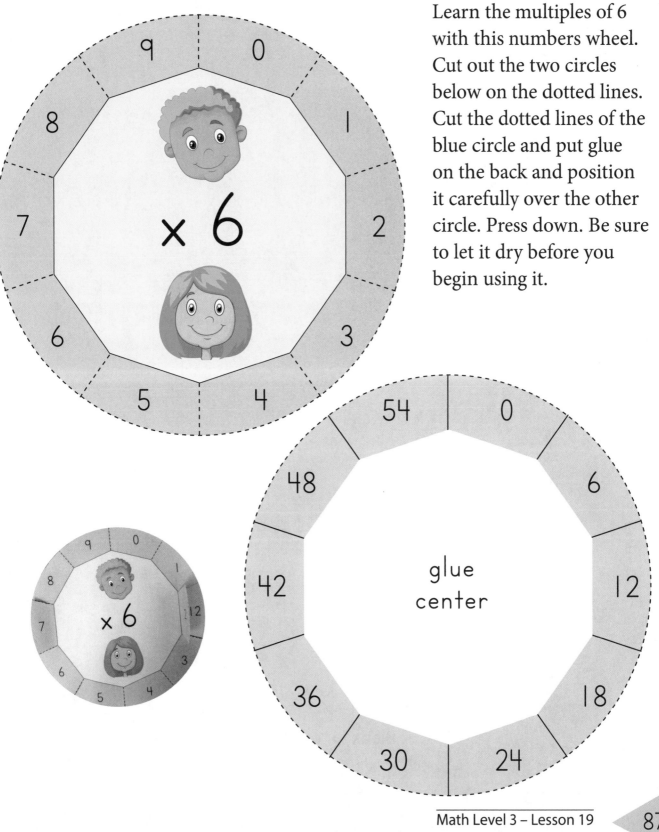

Blank for cutting.

Name_____

Introducing Multiplying and Dividing by 8 and 9

Multiply by 8 (fill in the blanks).

8   ____   ____   ____   ____   ____   ____   ____   ____   80

(8x1)   (8x2)   (8x3)   (8x4)   (8x5)   (8x6)   (8x7)   (8x8)   (8x9)   (8x10)

Circle groups of 8.

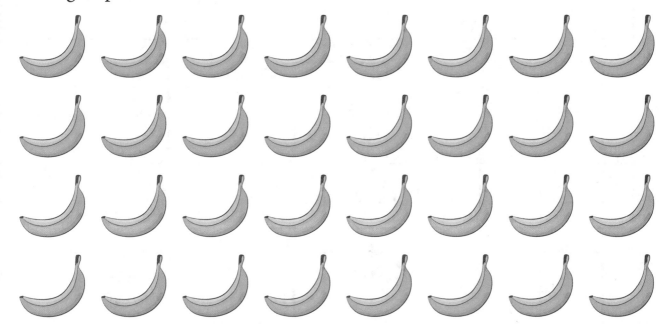

Answer these questions.

How many bananas are there? _____

How many groups did you circle? _____

How many are in each group? _____

Name_____

Missing Multiplication Number

Write the missing multiplication number in the box. The first one is done for you.

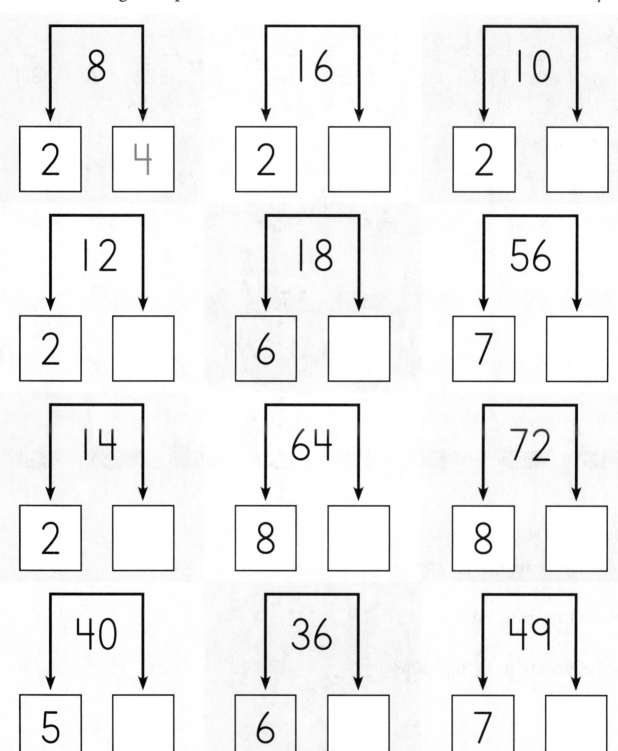

Name_____

Cut out the 9x study-buddy and use it to practice multiplying by 9.

	0×9
	1×9
	2×9
	3×9
	4×9
	5×9
	6×9
	7×9
	8×9
	9×9
	10×9

	9
	9 + 9
	9 + 9 + 9
	9 + 9 + 9 + 9
	9 + 9 + 9 + 9 + 9
	9 + 9 + 9 + 9 + 9 + 9
	9 + 9 + 9 + 9 + 9 + 9 + 9
	9 + 9 + 9 + 9 + 9 + 9 + 9 + 9
	9 + 9 + 9 + 9 + 9 + 9 + 9 + 9 + 9
	9 + 9 + 9 + 9 + 9 + 9 + 9 + 9 + 9 + 9

Introducing Rounding to 1000s and Estimation

Place Value Review

Use your Place Value Slider to show your teacher these numbers.

5,612   7,810   9,672   3,991

Rounding

Round these to the nearest 10.

28 _____   43 _____

99 _____   51 _____

Challenge! 167 _____

Round these to the nearest 100.

315 _____   873 _____

923 _____   231 _____

Challenge! 3,410 _____

Name_____

Critical Thinking

When the storm knocked the electricity out, Mom helped Dad light 49 candles. Charlie, Hairo, and Charlotte worked together to set up 52 cots. Estimate how many candles and cots there were at the children's home.

Round (to the nearest 10) the number of candles. _____

Round (to the nearest 10) the number of cots. _____

Estimate how many there were all together. _____ (cots and candles)

Name_____

Missing dots. Fill in the missing dots to make the addition equation true.

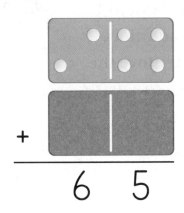

+

6   5

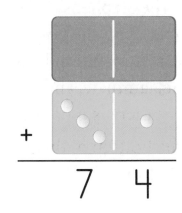

+

7   4

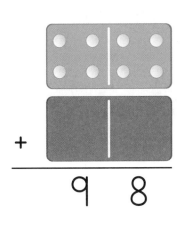

+

9   8

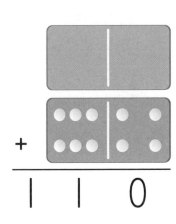

+

1 1   0

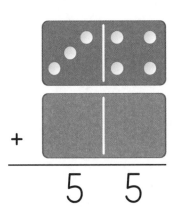

+

5   5

Practice Through Play

Use The Stephens' Trip to the Carnival Game (blank game board template) on the next spread to create a math equation (addition, subtraction, multiplication, and/or division) game. Just remove the game, laminate, and write math problems on all of the game squares except where there is mud. Use the instructions on the next page.

At the bottom of this page are the game pieces that you cut out and glue to bottle caps.

game pieces

The Stephens' Trip to the Carnival Game

Goal

Get the Stephens Family through the math maze to get to the merry-go-round.

Math Equation Race Game

To play the subtraction race game, you will need: two players (can be two students, one student and an older sibling, or a student and a parent), both of the game boards provided, two pencils, scrap paper for both players, the included game pieces, a die (one dice), a timer, and a calculator.

Steps:

Player 1 rolls the die and moves their game piece that many spaces.

If they land on a problem, they set the timer for 1.5 minutes and try solve the problem (the other player checks it on a calculator). If they get it correct, they get to go another turn. Limit three turns in a row for each player. If at any point in the game they land in a mud puddle, they have to go back to start.

Player 2 does the same thing.

To win, reach the finish line first.

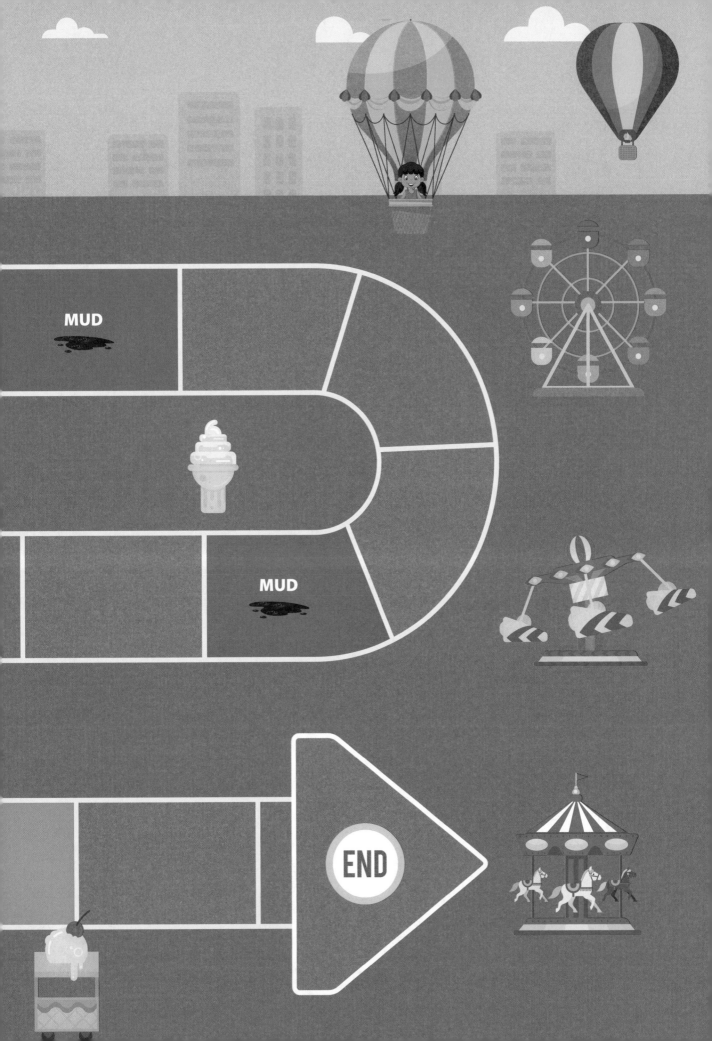

MUD

MUD

END

Introducing Higher Place Value through Millions

Multiplication Match-up

Draw a line from the multiplication problem to the addition problem it is equal to.

7 x 1	8 + 8
8 x 2	7 + 0
7 x 3	8 + 8 + 8 + 8
8 x 4	7 + 7 + 7
7 x 5	8 + 8 + 8 + 8 + 8 + 8
8 x 6	8 + 8 + 8
7 x 2	7 + 7 + 7 + 7 + 7
8 x 3	7 + 7
7 x 4	8 + 8 + 8 + 8 + 8
8 x 5	7 + 7 + 7 + 7

Mutt Math: Division

Divide the numbers on the outside of the wheel by the number in the center.
Write your answer in the outer ring. The first one is done for you.

Place Value Practice and Exploration

278 = _____ hundreds _____ tens _____ ones

1,629 = _____ thousands _____ hundreds _____ tens _____ ones

70 = _____ tens _____ ones

7 hundreds 5 tens 8 ones = _____

4 tens 3 ones = _____

100 + 20 + 5 = _____

400 + 10 + 3 = _____

Use your Place Value Slider to show your teacher these numbers.

3,419   9,002   5,555   7,126

Name_____

Working with Tally Marks and Odds and Evens

Count the tally marks and write how many.

卌 卌 卌 卌 卌 IIII   _____

卌 卌 卌 卌 卌 卌 卌 卌 II   _____

卌 卌 卌 卌 卌 卌 卌 I   _____

Draw each number in tally marks.

22

37

41

More Measurement Concepts

Measurement Match-up!

For each measurement below, circle whether it is weight, length, volume, or time.

1 pound = 16 ounces length weight	1 yard = 3 feet volume length
1 minute = 60 seconds time volume	1 pint = 2 cups length volume
1 hour = 60 minutes weight time	1 foot = 12 inches weight length
1 day = 24 hours time weight	1 quart = 2 pints time volume
1 gallon = 4 quarts volume length	1 year = 12 months time volume

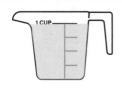

Rounding

Round these numbers to the nearest 10.

74 _____   127 _____

18 _____   144 _____

Round these numbers to the nearest 100.

345 _____   230 _____

792 _____   965 _____

Round these numbers to the nearest 1,000.

3,451 _____   8,900 _____

1,642 _____   19,203 _____

Placing Commas

Place commas in the correct place in these numbers. Read them out loud to your teacher.

381023   12304

8910246   20394356

Name_____

Mixed Review

Add and Subtract. Watch the signs.

$$
\begin{array}{r}
23,875 \\
-\ 2,945 \\
\hline
\end{array}
\qquad
\begin{array}{r}
18,237 \\
-\ 6,125 \\
\hline
\end{array}
\qquad
\begin{array}{r}
34,986 \\
-\ 3,893 \\
\hline
\end{array}
$$

$$
\begin{array}{r}
264 \\
321 \\
+\ \ \ 87 \\
\hline
\end{array}
\qquad
\begin{array}{r}
4,820 \\
231 \\
+\ \ \ 54 \\
\hline
\end{array}
\qquad
\begin{array}{r}
5,930 \\
1,201 \\
+\ \ \ 334 \\
\hline
\end{array}
$$

Finding Value

Write the value of the underlined number. The first one is done for you.

4,4<u>4</u>5   <u>7</u>,439   70,3<u>8</u>1   8<u>1</u>6,511   <u>4</u>,321,792

400   _____   _____   _____   _____

Name_____

Measurement Multiple Choice

Circle whether the object should be measured in inches, feet, yards, or miles.

ticket		tomato	
	inches		inches
	yards		miles

barn		cat	
	inches		inches
	yards		yards

earth		highway	
	yards		feet
	miles		miles

flag		table	
	inches		feet
	feet		yards

Name_____

Multiplication Grids

Shade each grid to match the multiplication fact and write in the missing answer. The first one is done for you.

3 x 5 = 15

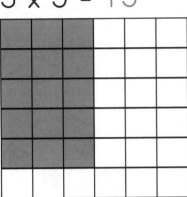

4 x 4 =

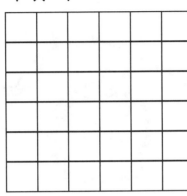

6 x 1 =

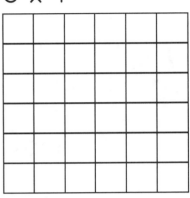

6 x 2 =

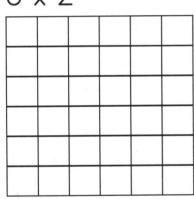

3 x 3 =

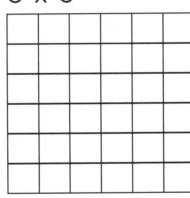

5 x 5 =

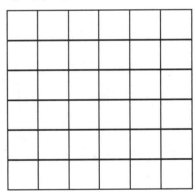

4 x 6 =

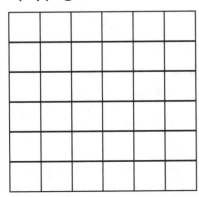

6 x 6 =

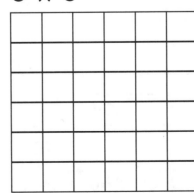

6 x 3 =

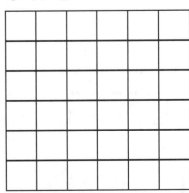

Name_____

Solve!

Solve the division problems to find out the correct colors.

$16 \div 4 =$

$14 \div 2 =$

$20 \div 5 =$

$30 \div 6 =$

$21 \div 3 =$

$24 \div 4 =$

$36 \div 6 =$

$49 \div 7 =$

$12 \div 2 =$

$18 \div 3 =$

$12 \div 4 =$

$18 \div 6 =$

$49 \div 7 =$

$35 \div 5 =$

$25 \div 5 =$

$30 \div 10 =$

$8 \div 4 =$

$8 \div 2 =$

Blue ▷ 5

Grey ▷ 6

Orange ▷ 2

Green ▷ 3

Yellow ▷ 4

Red ▷ 7

Introducing Inequalities

Greater Than/Less Than. Circle the Greater Than Croc if the first number in each box is greater than the second. Circle the Less Than Croc if the first number is less. Remember, the wider side is the larger numbered side.

 < = Less Than   Greater Than = >

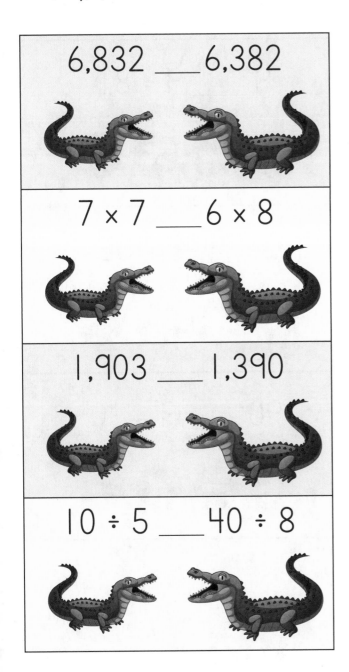

6,832 ___ 6,382

7 × 7 ___ 6 × 8

1,903 ___ 1,390

10 ÷ 5 ___ 40 ÷ 8

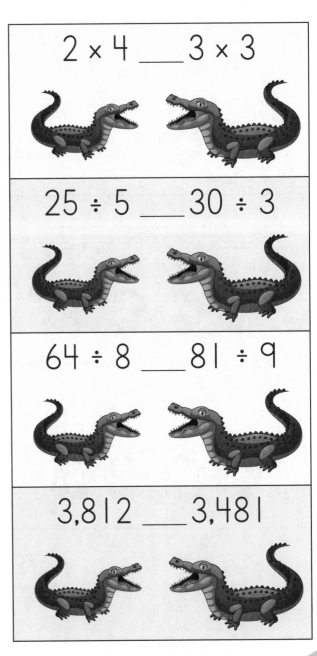

2 × 4 ___ 3 × 3

25 ÷ 5 ___ 30 ÷ 3

64 ÷ 8 ___ 81 ÷ 9

3,812 ___ 3,481

Find the Sign!

In each box, circle the sign that makes the statement true.

10 ___ 4 = 14 	20 ___ 5 = 4
4 ___ 8 = 12 	9 ___ 9 = 81
20 ___ 8 = 12 	7 ___ 8 = 56
15 ___ 5 = 20 	90 ___ 9 = 10

Multiplication with Missing Factors

Fill in the blank with the missing factor to make the equation true.

6 x ___ = 36

3 x ___ = 27

8 x ___ = 40

9 x ___ = 81

7 x ___ = 56

2 x ___ = 20

5 x ___ = 45

4 x ___ = 36

3 x ___ = 15

9 x ___ = 27

8 x ___ = 64

7 x ___ = 63

4 x ___ = 20

2 x ___ = 24

6 x ___ = 36

3 x ___ = 30

5 x ___ = 30

8 x ___ = 72

10 x ___ = 100

10 x ___ = 60

Find the Sign!

Charlotte loves to splash paint. She has painted equal and not equal signs.
Circle the sign that she has painted to make the statement in each box true.

3 + 8 ___ 7 + 5 = ≠	30 ÷ 10 ___ 9 ÷ 3 = ≠
6 + 7 ___ 3 + 9 = ≠	25 ÷ 5 ___ 50 ÷ 10 = ≠

She has now painted greater than and less than signs. Circle the sign that she
has painted to make the statement in each box true.

36 ÷ 4 ___ 3 × 5 < >	20 ÷ 5 ___ 6 + 2 < >
2 × 9 ___ 7 + 4 < >	100 ÷ 10 ___ 4 × 2 < >

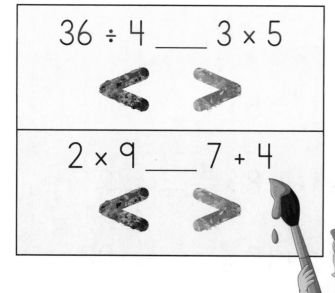
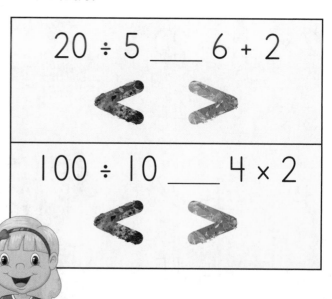

Review of New Concepts

Round these numbers to the nearest 100.

239 _____   175 _____

967 _____   428 _____

201 _____   891 _____

Round these numbers to the nearest 1,000.

3,560 _____   2,910 _____

4,239 _____   8,143 _____

5,756 _____   19,345 _____

Name_____

Rounding

The children decided to dress up in costumes. Help superhero Lightning Nat save the world by rounding these to the nearest 100 and then estimate the sum.

$$589$$
$$+ \ 230 \longrightarrow$$
$$+ \ _____$$

$$207$$
$$+ \ 333 \longrightarrow$$
$$+ \ _____$$

Round these to the nearest 1,000 and then estimate the sum.

$$2,349$$
$$+ \ 1,259 \longrightarrow$$
$$+ \ _____$$

$$4,789$$
$$+ \ 5,012 \longrightarrow$$
$$+ \ _____$$

Name_____

Missing Division Number

Write the missing division number in the box. The first one is done for you.

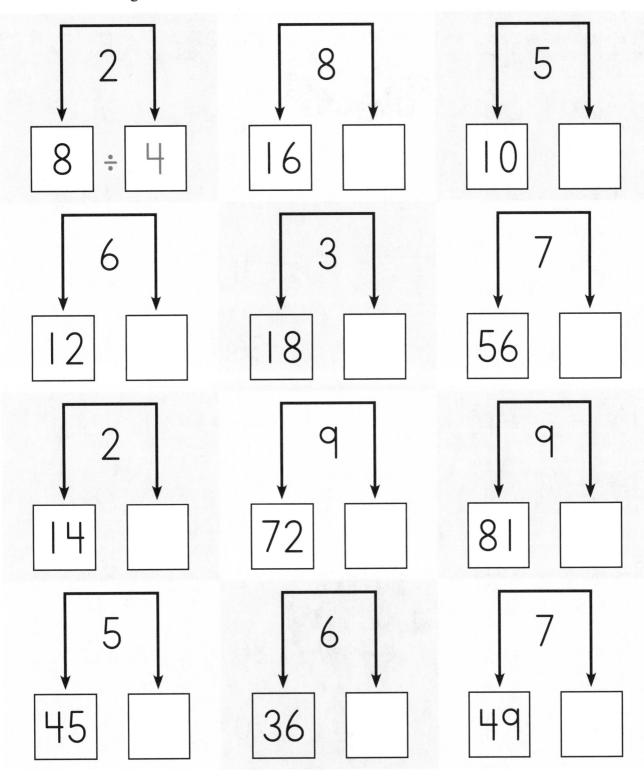

Name_____

Division with Missing Factors

Fill in the blank with the missing factor to make the equation true.

$36 \div$ ___ $= 6$

$27 \div$ ___ $= 3$

$40 \div$ ___ $= 8$

$81 \div$ ___ $= 9$

$56 \div$ ___ $= 56$

$20 \div$ ___ $= 10$

$45 \div$ ___ $= 9$

$36 \div$ ___ $= 4$

$15 \div$ ___ $= 3$

$27 \div$ ___ $= 9$

$64 \div$ ___ $= 8$

$63 \div$ ___ $= 7$

$20 \div$ ___ $= 10$

$24 \div$ ___ $= 2$

$36 \div$ ___ $= 6$

$30 \div$ ___ $= 15$

$30 \div$ ___ $= 6$

$72 \div$ ___ $= 8$

$80 \div$ ___ $= 10$

$70 \div$ ___ $= 7$

Addition and Subtraction of Larger Numbers

Solve these addition problems. Be sure to place the comma correctly in your answers.

$$
\begin{array}{r}
1{,}368 \\
+\ 2{,}665 \\
\hline
\end{array}
\qquad
\begin{array}{r}
5{,}580 \\
+\ 1{,}258 \\
\hline
\end{array}
\qquad
\begin{array}{r}
2{,}827 \\
+\ 4{,}350 \\
\hline
\end{array}
$$

$$
\begin{array}{r}
3{,}146 \\
+\ 4{,}887 \\
\hline
\end{array}
\qquad
\begin{array}{r}
6{,}703 \\
+\ 2{,}401 \\
\hline
\end{array}
\qquad
\begin{array}{r}
2{,}604 \\
+\ 6{,}572 \\
\hline
\end{array}
$$

$$
\begin{array}{r}
5{,}224 \\
+\ 7{,}448 \\
\hline
\end{array}
\qquad
\begin{array}{r}
8{,}521 \\
+\ 3{,}477 \\
\hline
\end{array}
\qquad
\begin{array}{r}
7{,}506 \\
+\ 4{,}275 \\
\hline
\end{array}
$$

$$
\begin{array}{r}
6{,}826 \\
+\ 5{,}779 \\
\hline
\end{array}
\qquad
\begin{array}{r}
9{,}800 \\
+\ 2{,}700 \\
\hline
\end{array}
\qquad
\begin{array}{r}
9{,}999 \\
+\ 4{,}444 \\
\hline
\end{array}
$$

Name_____

Solve!

Solve these addition problems. Be sure to place the comma correctly in your answers.

$$\begin{array}{r} 11,278 \\ +\ 10,465 \\ \hline \end{array} \qquad \begin{array}{r} 15,850 \\ +\ 11,258 \\ \hline \end{array} \qquad \begin{array}{r} 18,727 \\ +\ 14,354 \\ \hline \end{array}$$

Mental Math. Be sure to tell your teacher the answers.

$$12 + 2 - 4 + 6 = \qquad 3 \times 9 =$$

$$81 \div 9 = \qquad 3 + 8 + 16 - 15 =$$

Write +, −, **x**, or ÷.

$$4 \_\_ 3 = 12 \qquad 13 \_\_ 13 = 0 \qquad 9 \_\_ 9 = 18$$

Name_____

Solve!

Solve these subtraction problems. Be sure to place the comma correctly in your answers.

$$\begin{array}{r} 6,345 \\ - 4,131 \\ \hline \end{array}$$
$$\begin{array}{r} 7,178 \\ - 5,045 \\ \hline \end{array}$$
$$\begin{array}{r} 6,842 \\ - 6,343 \\ \hline \end{array}$$

$$\begin{array}{r} 7,234 \\ - 3,244 \\ \hline \end{array}$$
$$\begin{array}{r} 8,056 \\ - 6,158 \\ \hline \end{array}$$
$$\begin{array}{r} 3,344 \\ - 2,188 \\ \hline \end{array}$$

$$\begin{array}{r} 2,467 \\ - 1,468 \\ \hline \end{array}$$
$$\begin{array}{r} 6,168 \\ - 3,269 \\ \hline \end{array}$$
$$\begin{array}{r} 8,765 \\ - 5,679 \\ \hline \end{array}$$

$$\begin{array}{r} 5,734 \\ - 4,789 \\ \hline \end{array}$$
$$\begin{array}{r} 7,057 \\ - 4,371 \\ \hline \end{array}$$
$$\begin{array}{r} 9,873 \\ - 6,649 \\ \hline \end{array}$$

Clock and Time Concepts Practice

Fill in the blank with what the time will be and then, draw the hands in each clock showing the correct time.

Right now it is 3:15, in 45 minutes, it will be

_____ : _____

Right now it is 7:30, in one hour and 10 minutes, it will be

_____ : _____

Right now it is 12:50, in one hour and 20 minutes, it will be

_____ : _____

Right now it is 6:55, in 50 minutes, it will be

_____ : _____

Right now it is 9:00, in 2 hours and 15 minutes, it will be

_____ : _____

Name_____

Introducing Roman Numerals

Match the Roman numerals with the standard numbers.

V   100

C   50

D   1,000

L   10

I   500

X   1

M   5

Copywork of Roman Numerals

I	II	III	IV	V

VI	VII	VIII	IX	X

XI	XII	XIII	XIV	XV

XVI	XVII	XVIII	XIX	XX

XXI	XXII	XXIII	XXIV	XXV

XXVI	XXVII	XXVIII	XXIX	XXX

Name_____

Multiplication

Practice multiplication facts. Fill in the chart with the correct numbers.

×	1	2	3	4	5	6	7	8	9	10
1										
2										
3										
4										
5										

Roman Numerals

Convert these numbers into Roman numerals.

100 = _____   50 = _____

1,000 = _____   500 = _____

3 = _____   12 = _____

7 = _____   21 = _____

9 = _____   30 = _____

Name_____

What's the number? Fill in the blanks with the standard number.

C = _____   L = _____

M = _____   D = _____

IV = _____   VII = _____

IX = _____   III = _____

XX = _____   XVI = _____

LX = _____   XXI = _____

Multiplication

Practice multiplication facts. Fill in the chart with the correct numbers.

×	1	2	3	4	5	6	7	8	9	10
6										
7										
8										
9										
10										

More about Roman Numerals

Solve the following problems. Be sure to check the signs to see if it is addition or subtraction.

```
    5,320          7,102          8,201
    1,264          2,385          9,642
 + 4,800        + 5,173        + 6,381
```

```
      933            821            667
      812            705            443
      755            623            559
    + 671          + 414          + 756
```

```
    7,740          8,267          9,456
  - 4,954        - 7,612        - 8,673
```

Money

Using play money, count out the following amounts.

$8.73 $ .82

$10.55 $5.19

Now work through these problems. Be sure to write your answers correctly with commas, dollar signs, or cent signs when needed.

If an item cost $2.20, and you paid with a $5 bill, what would be your change?

If an item cost $9.62, and you paid with a $10 bill, what would be your change?

If your total at the store was $62.38, and you paid with a $100 bill, what would be your change?

Name_____

Color by Roman Numeral

Color the image below using Roman numerals to determine the colors.

Name_____

Match the equation. Draw a line to match the equation or Roman numeral on the left with the matching number on the right.

6 × 6	14
XX	18
IV	8
40 ÷ 4	25
9 × 8	9
II	19
XVIII	36
10 + 4	20
20 – 6	10
5 × 5	4
XIX	72
64 ÷ 8	14
81 ÷ 9	2

Quiz Section

Name_____

Add or Subtract

Watch the signs! Circle the problems where you used carrying and borrowing.

$$\begin{array}{r} 845 \\ -\ 762 \\ \hline \end{array}$$

$$\begin{array}{r} 631 \\ -\ 430 \\ \hline \end{array}$$

$$\begin{array}{r} 285 \\ +\ 102 \\ \hline \end{array}$$

$$\begin{array}{r} 429 \\ -\ 379 \\ \hline \end{array}$$

$$\begin{array}{r} 45 \\ 23 \\ +\ 15 \\ \hline \end{array}$$

$$\begin{array}{r} 812 \\ 645 \\ +\ 237 \\ \hline \end{array}$$

Mental Math

Read these numbers to your teacher.

358   634

344   999

Measurement. Draw lines starting at the star.

2 inches *

$5\frac{1}{2}$ inches *

$4\frac{1}{2}$ inches *

Time

Fill in the clock faces to show the correct time.

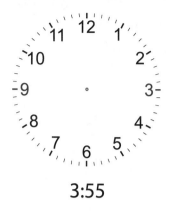 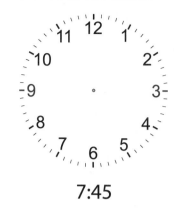 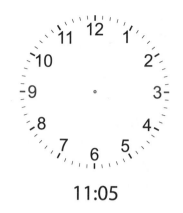

3:55   7:45   11:05

Rounding

Round these numbers to the nearest 100.

346 _____

    Which hundreds is it between? _____ and _____

    Look at the digit in the tens' place. Is it 5 or greater? _____

591 _____

    Which hundreds is it between? _____ and _____

    Look at the digit in the tens' place. Is it 5 or greater? _____

723 _____

    Which hundreds is it between? _____ and _____

    Look at the digit in the tens' place. Is it 5 or greater? _____

Fractions

Write **T** for true or **F** for false.

_____ The bottom number in a fraction is called the numerator.

_____ If the numerator in a fraction is 1, that means we are talking about 1 part of the whole.

_____ In the fractions $\frac{1}{10}$ and $\frac{1}{5}$, the second fraction represents a smaller piece of the whole.

_____ The top number in a fraction is called the numerator.

Circle groups of 5 squares and answer the questions.

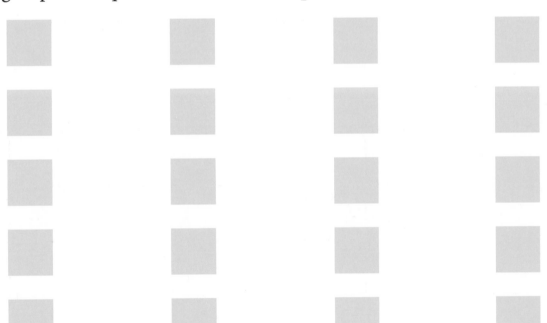

How many groups? _____   How many in each group? _____

$4 \times 5 =$ _____ $\frac{1}{4}$ of 20 is _____

Perimeter and Area

If you had a square that had sides that were 3 feet long, what would be the perimeter? What would be the area?

Perimeter: _____ x _____ = _____ feet

Area: _____ x _____ = _____ square feet

Multiplication. Fill in the multiplication grid.

×	1	2	3	4	5	6	7	8	9	10
1										
2										
3										
4										
5										
6										
7										
8										
9										
10										100

Divide

$30 \div 3 =$

$27 \div 3 =$

$15 \div 3 =$

$25 \div 5 =$

$20 \div 2 =$

$36 \div 4 =$

$12 \div 4 =$

$40 \div 10 =$

Critical Thinking

Show your teacher at least one multiplication fact to go with each of the division facts above. For example:

$3 \times 10 = 30$

$10 \times 3 = 30$

Add or Subtract

$$\begin{array}{r} 562 \\ -\ 388 \\ \hline \end{array}$$

$$\begin{array}{r} 742 \\ -\ 540 \\ \hline \end{array}$$

$$\begin{array}{r} 396 \\ +\ 204 \\ \hline \end{array}$$

$$\begin{array}{r} 4{,}501 \\ +\ \ \ 320 \\ \hline \end{array}$$

$$\begin{array}{r} 2{,}912 \\ -\ 1{,}890 \\ \hline \end{array}$$

$$\begin{array}{r} 341 \\ 55 \\ +\ \ 23 \\ \hline \end{array}$$

Place Value

Fill in the blanks.

35,120

In this number,

3 = ____ groups of _____.

5 = ____ groups of _____.

1 = ____ groups of _____.

2 = ____ groups of _____.

0 = ____ groups of ____.

5,623,890

In this number,

5 = ____ groups of _____.

6 = ____ groups of _____.

2 = ____ groups of _____.

3 = ____ groups of _____.

8 = ____ groups of _____.

9 = ____ groups of _____.

0 = ____ groups of ____.

What's the number? Read these numbers to your teacher.

32,678,103   10,592,682

768,012

What's the sign? Fill in > or <.

$\frac{1}{4}$ ___ $\frac{1}{8}$

$12 \div 3$ ___ $15 \div 5$

18 ___ 81

What's the sign? Fill in = or ≠.

24 ÷ 6 ___ 16 ÷ 4

64 ÷ 8 ___ 27 ÷ 3

1 ton ___ 2,000 pounds

1 mile ___ 5,280 feet

1,760 yards ___ 1 mile

7 feet ___ 2 yards

Solve. Solve these story problems.

Mom bought the children a treat after their walk around Lima's market. The treat cost $8.76. If she paid with a $10 bill, how much was her change?

The local churches raised $450 to go towards building the clinic at the children's home. The electrician's work cost $389. How much was left over after paying the electrician for his work?

Roman Numerals

Next to each standard numeral, write the matching Roman numeral.

1 _____	11 _____
2 _____	12 _____
3 _____	13 _____
4 _____	14 _____
5 _____	15 _____
6 _____	16 _____
7 _____	17 _____
8 _____	18 _____
9 _____	19 _____
10 _____	20 _____

Multiply and Divide

Multiply each number by 7.

8	10	6	5	7	9

Divide each number by 8.

80	64	48	32	16	8

Find the area and perimeter of each of these shapes.

A square with 5-foot sides.

Area: _____

Perimeter: _____

A rectangle with two 7-inch sides and two 4-inch sides.

Area: _____

Perimeter: _____

Divide

Divide each of these shapes to match the fractions below them. Circle the largest fractional piece.

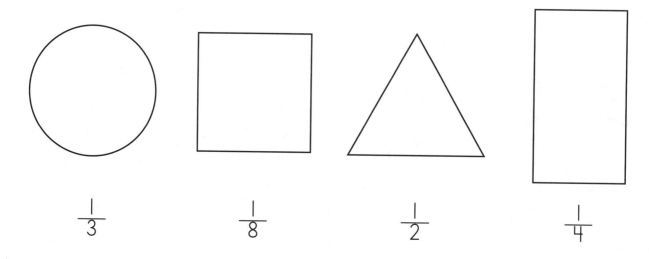

$$\frac{1}{3} \qquad \frac{1}{8} \qquad \frac{1}{2} \qquad \frac{1}{4}$$

Solutions Manual: Lessons 1 – 2

Name_____

Evens or Odds? Circle whether the number is odd or even. Even numbers always end in 0, 2, 4, 6, 8. Odd numbers always end in 1, 3, 5, 7, 9.

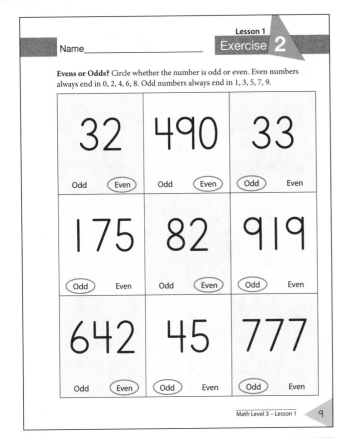

32	490	33
Odd (Even)	Odd (Even)	(Odd) Even
175	82	919
(Odd) Even	Odd (Even)	(Odd) Even
642	45	777
Odd (Even)	(Odd) Even	(Odd) Even

Name_____

Dot-to-Dot. Starting with the number 1, connect the dots up to number 56. Then color your picture.

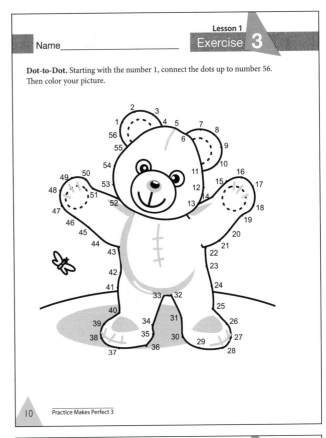

Name_____

Review of Money, Clocks, Perimeter, Addition/Subtraction Facts

Which clock is it? Circle the correct clock.

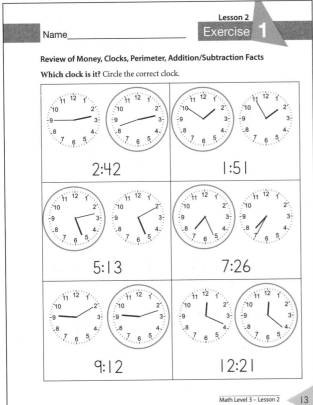

Name_____

Welcome to the Amusement Park!
Figure out your budget for the carnival rides below. The first one is done for you.

Solutions Manual: Lessons 2 – 3

What time is it? Write the correct time in each box.

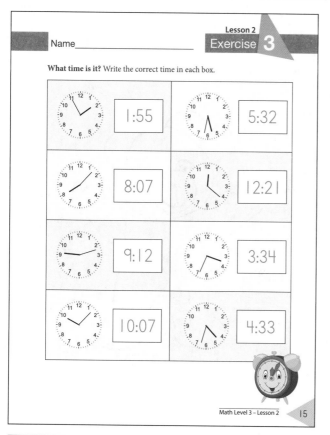

1:55   5:32

8:07   12:21

9:12   3:34

10:07   4:33

Perimeter in My Room

Find the perimeter of these objects found in a child's room. The first one is done for you.

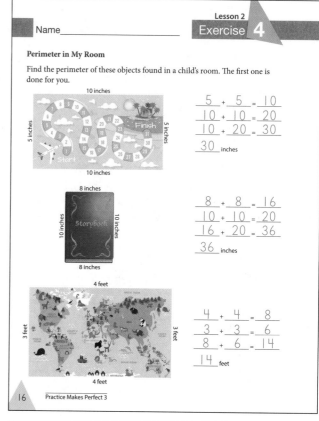

$5 + 5 = 10$
$10 + 10 = 20$
$10 + 20 = 30$
30 inches

$8 + 8 = 16$
$10 + 10 = 20$
$16 + 20 = 36$
36 inches

$4 + 4 = 8$
$3 + 3 = 6$
$8 + 6 = 14$
14 feet

Review of Addition, Including Carrying, Tally Marks

Missing dots. Draw in the missing dots to make the addition equation true. The first one is done for you.

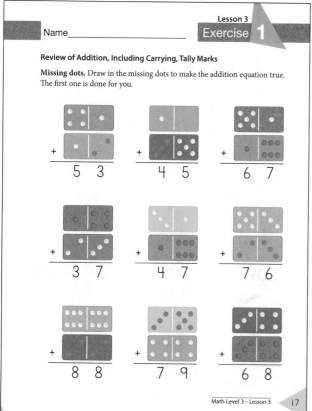

   5   3   4   5   6   7

   3   7   4   7   7   6

   8   8   7   9   6   8

Skip Counting by 5

Add 5 to the last blue bubble and color it blue to help the turtle find his way to the starfish at the end of the maze. Examples: $5 + 5 = 10$, $10 + 5 = 15$

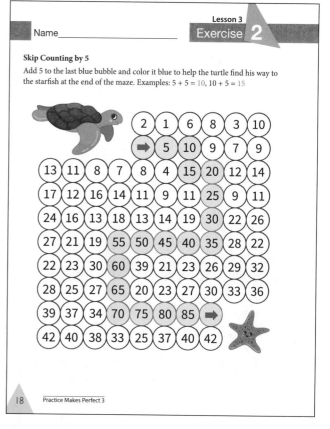

Solutions Manual: Lessons 3 – 4

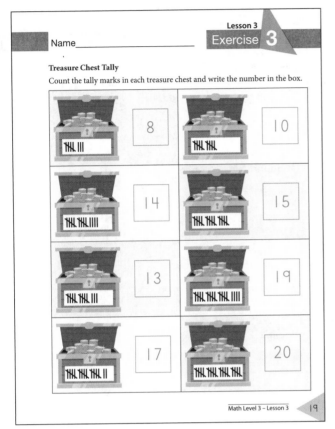

Exercise 3

Name_____

Treasure Chest Tally
Count the tally marks in each treasure chest and write the number in the box.

𝖳𝖧𝖫 III	8
𝖳𝖧𝖫 𝖳𝖧𝖫	10
𝖳𝖧𝖫 𝖳𝖧𝖫 IIII	14
𝖳𝖧𝖫 𝖳𝖧𝖫 𝖳𝖧𝖫	15
𝖳𝖧𝖫 𝖳𝖧𝖫 III	13
𝖳𝖧𝖫 𝖳𝖧𝖫 𝖳𝖧𝖫 IIII	19
𝖳𝖧𝖫 𝖳𝖧𝖫 𝖳𝖧𝖫 II	17
𝖳𝖧𝖫 𝖳𝖧𝖫 𝖳𝖧𝖫 𝖳𝖧𝖫	20

Math Level 3 – Lesson 3   19

Exercise 4

Name_____

Do the Math Challenge
Do the addition and subtraction to fill in this chart. The first one is done for you.

	1 More	1 Less	10 More	10 Less
12	13	11	22	2
18	19	17	28	8
15	16	14	25	5
20	21	19	30	10
16	17	15	26	6
40	41	39	50	30

20   Practice Makes Perfect 3

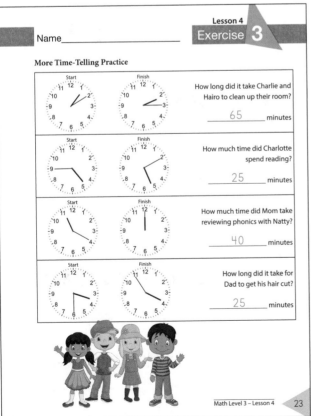

Exercise 3

Name_____

More Time-Telling Practice

How long did it take Charlie and Hairo to clean up their room?
_____65_____ minutes

How much time did Charlotte spend reading?
_____25_____ minutes

How much time did Mom take reviewing phonics with Natty?
_____40_____ minutes

How long did it take for Dad to get his hair cut?
_____25_____ minutes

Math Level 3 – Lesson 4   23

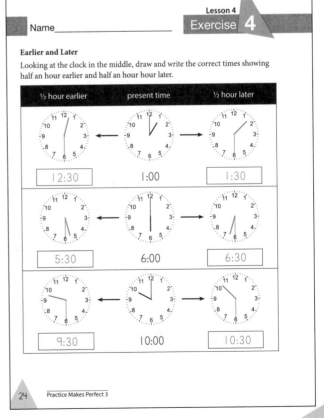

Exercise 4

Name_____

Earlier and Later
Looking at the clock in the middle, draw and write the correct times showing half an hour earlier and half an hour hour later.

½ hour earlier	present time	½ hour later
12:30	1:00	1:30
5:30	6:00	6:30
9:30	10:00	10:30

24   Practice Makes Perfect 3

Solutions Manual: Lesson 5

Name_____

Lesson 5
Exercise 1

Review of Measurement, Fractions, Thermometers, Graphs

Measure the Silverware!

Measure each item with the ruler provided below it. The first one is done for you.

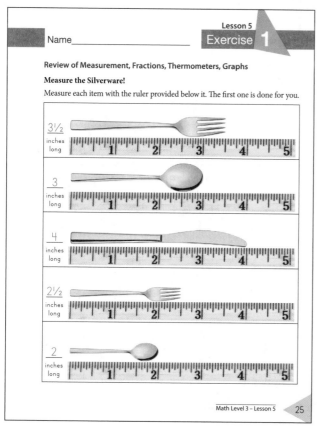

3½ inches long

3 inches long

4 inches long

2½ inches long

2 inches long

Name_____

Lesson 5
Exercise 2

Practice time!

Fill in the blank with all you remember.

16 ounces = __1__ pound

1 foot = __12__ inches

3 feet = __1__ yard

4 quarts = __1__ gallon

1 quart = __2__ pints

2 cups = __1__ pint

60 minutes = __1__ hour

12 months = __1__ year

60 seconds = __1__ minute

24 hours = __1__ day

1 week = __7__ days

Name_____

Lesson 5
Exercise 3

Count the Coins

Charlie has been practicing counting coins. Circle "thumbs up" if he gave the correct answer. Circle "thumbs down" if he gave the incorrect answer.

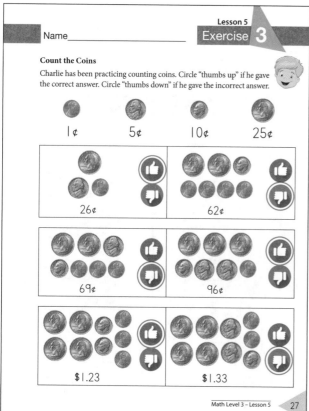

1¢   5¢   10¢   25¢

26¢   62¢

69¢   96¢

$1.23 $1.33

Name_____

Lesson 5
Exercise 4

Measuring with Cups, Pints, and Quarts

Look at the measurement chart to the right and then circle the correct amount below.

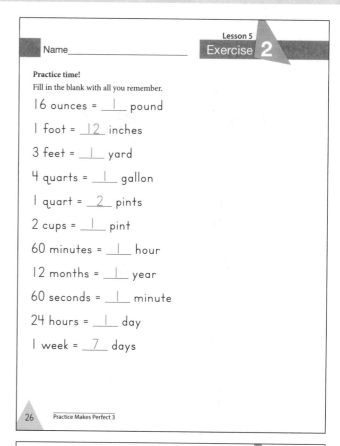
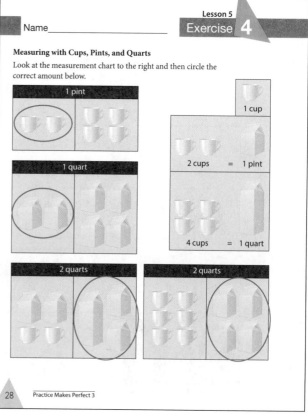

1 pint

1 quart

2 quarts   2 quarts

1 cup

2 cups = 1 pint

4 cups = 1 quart

Solutions Manual: Lessons 6 – 7

Name_____

Lesson 6
Exercise 1

Review of Word Problems

Since the entire lesson in your math book is centered around reviewing how to solve word problems, you may take some time each day and practice some other skills.

Ordering Numbers

Place the numbers in the correct order from smallest number to largest number.

332, 350, 312, 300	643, 421, 390, 199
<u>300</u>, <u>312</u>, <u>332</u>, <u>350</u>	<u>199</u>, <u>390</u>, <u>421</u>, <u>643</u>

454, 450, 452, 459	900, 867, 712, 910
<u>450</u>, <u>452</u>, <u>454</u>, <u>459</u>	<u>712</u>, <u>867</u>, <u>900</u>, <u>910</u>

231, 132, 321, 213	745, 457, 547, 754
<u>132</u>, <u>213</u>, <u>231</u>, <u>321</u>	<u>457</u>, <u>547</u>, <u>745</u>, <u>754</u>

101, 110, 220, 201	874, 478, 847, 487
<u>101</u>, <u>110</u>, <u>201</u>, <u>220</u>	<u>478</u>, <u>487</u>, <u>847</u>, <u>874</u>

Name_____

Lesson 6
Exercise 2

Yesterday, Today, and Tomorrow. Write down the day before and the day after for each day in the "Today" column.

Yesterday	Today	Tomorrow
Wednesday	**Thursday**	Friday
Monday	**Tuesday**	Wednesday
Friday	**Saturday**	Sunday
Saturday	**Sunday**	Monday
Thursday	**Friday**	Saturday
Tuesday	**Wednesday**	Thursday
Sunday	**Monday**	Tuesday

Name_____

Lesson 7
Exercise 1

Introducing Column Addition and Adding Larger Numbers

Vertical Addition

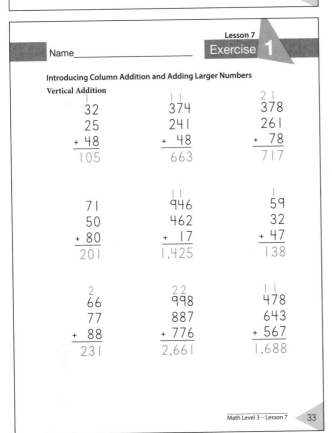

$$\begin{array}{r} 32 \\ 25 \\ +48 \\ \hline 105 \end{array} \qquad \begin{array}{r} 374 \\ 241 \\ +48 \\ \hline 663 \end{array} \qquad \begin{array}{r} 378 \\ 261 \\ +78 \\ \hline 717 \end{array}$$

$$\begin{array}{r} 71 \\ 50 \\ +80 \\ \hline 201 \end{array} \qquad \begin{array}{r} 946 \\ 462 \\ +17 \\ \hline 1,425 \end{array} \qquad \begin{array}{r} 59 \\ 32 \\ +47 \\ \hline 138 \end{array}$$

$$\begin{array}{r} 66 \\ 77 \\ +88 \\ \hline 231 \end{array} \qquad \begin{array}{r} 998 \\ 887 \\ +776 \\ \hline 2,661 \end{array} \qquad \begin{array}{r} 478 \\ 643 \\ +567 \\ \hline 1,688 \end{array}$$

Name_____

Lesson 7
Exercise 2

Counting Change. Add up the amount of each set of coins and write down your answer.

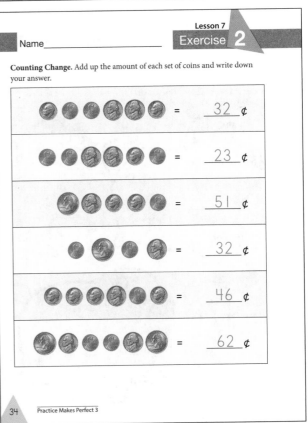

= <u>32</u> ¢

= <u>23</u> ¢

= <u>51</u> ¢

= <u>32</u> ¢

= <u>46</u> ¢

= <u>62</u> ¢

Solutions Manual: Lessons 7 – 8

Name_____

Plus or Minus? Decide which operation: Addition or Subtraction? Fill in the blanks with a + or a −.

6 _+_ 3 = 9   16 _−_ 3 = 13

8 _−_ 3 = 5   15 _+_ 3 = 18

8 _+_ 4 = 12   10 _+_ 8 = 18

9 _−_ 3 = 6   10 _+_ 2 = 12

7 _−_ 4 = 3   11 _−_ 4 = 7

3 _+_ 8 = 11   18 _−_ 9 = 9

4 _+_ 2 = 6   10 _−_ 5 = 5

Math Level 3 – Lesson 7   35

Name_____

Place Value Practice

Write down these numbers by using your **Place Value Slider** which you made in Lesson 1 of these *Practice Makes Perfect* worksheets.

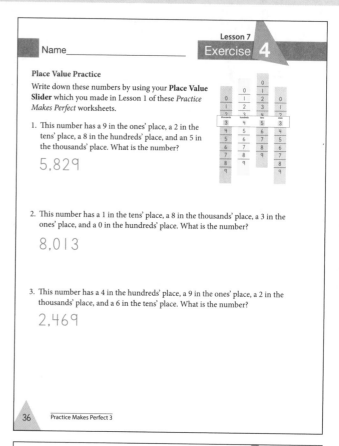

1. This number has a 9 in the ones' place, a 2 in the tens' place, a 8 in the hundreds' place, and an 5 in the thousands' place. What is the number?

   5,829

2. This number has a 1 in the tens' place, a 8 in the thousands' place, a 3 in the ones' place, and a 0 in the hundreds' place. What is the number?

   8,013

3. This number has a 4 in the hundreds' place, a 9 in the ones' place, a 2 in the thousands' place, and a 6 in the tens' place. What is the number?

   2,469

36   Practice Makes Perfect 3

Name_____

Introducing Larger Number Subtraction

Mutt Math: Addition

Today, let's practice addition facts. Add each number on the outside of the wheel to the number in the center and write the answer.

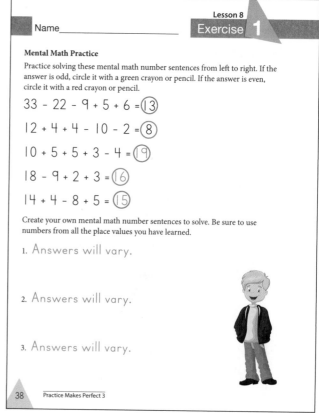

Math Level 3 – Lesson 8   37

Name_____

Mental Math Practice

Practice solving these mental math number sentences from left to right. If the answer is odd, circle it with a green crayon or pencil. If the answer is even, circle it with a red crayon or pencil.

33 − 22 − 9 + 5 + 6 = ⑬

12 + 4 + 4 − 10 − 2 = ⑧

10 + 5 + 5 + 3 − 4 = ⑲

18 − 9 + 2 + 3 = ⑯

14 + 4 − 8 + 5 = ⑮

Create your own mental math number sentences to solve. Be sure to use numbers from all the place values you have learned.

1. Answers will vary.

2. Answers will vary.

3. Answers will vary.

38   Practice Makes Perfect 3

Solutions Manual: Lesson 8

Mutt Math: Subtraction

Today, let's practice subtraction facts. Subtract the numbers around the middle of the wheel from the number in the center. Write the answer on the outside ring.

Calendar Practice

For each question below, draw a line to the correct month.

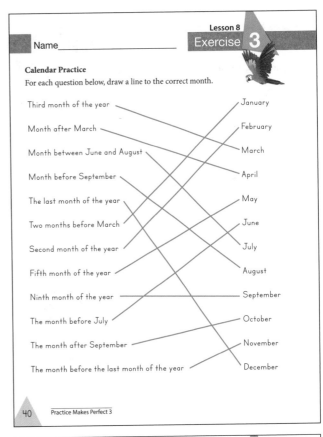

Third month of the year

Month after March

Month between June and August

Month before September

The last month of the year

Two months before March

Second month of the year

Fifth month of the year

Ninth month of the year

The month before July

The month after September

The month before the last month of the year

January

February

March

April

May

June

July

August

September

October

November

December

How long and how much?

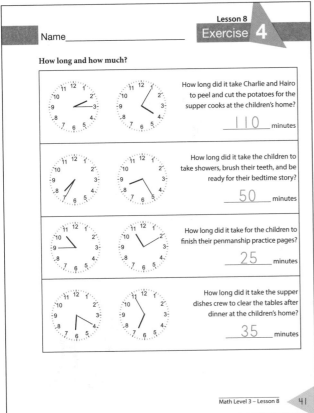

How long did it take Charlie and Hairo to peel and cut the potatoes for the supper cooks at the children's home?

____1 1 0____ minutes

How long did it take the children to take showers, brush their teeth, and be ready for their bedtime story?

____50____ minutes

How long did it take for the children to finish their penmanship practice pages?

____25____ minutes

How long did it take the supper dishes crew to clear the tables after dinner at the children's home?

____35____ minutes

Odds and Evens

Circle whether the number is odd or even.

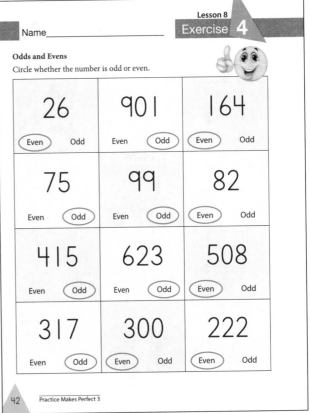

26	901	164
(Even) Odd	Even (Odd)	(Even) Odd
75	99	82
Even (Odd)	Even (Odd)	(Even) Odd
415	623	508
Even (Odd)	Even (Odd)	(Even) Odd
317	300	222
Even (Odd)	(Even) Odd	(Even) Odd

Solutions Manual: Lesson 9

Name_____

Lesson 9
Exercise 1

Introducing Rounding to the 10s and 100s

Follow these steps to understand the idea of rounding.

Step 1: Write or think which tens the number is in between.
Step 2: Look at the digit in the ones' place. Is it greater than or less than 5?
Step 3: If the digit in the ones' place is 5 or greater than 5, round to the larger ten.
Step 4: If the digit in the ones' place is less than 5, round to the smaller ten.

Show your teacher how to round these numbers to the nearest 10. Circle the correct answer.

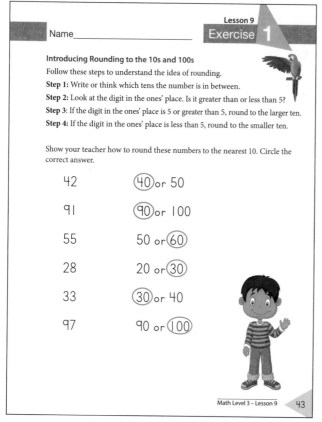

42	(40) or 50
91	(90) or 100
55	50 or (60)
28	20 or (30)
33	(30) or 40
97	90 or (100)

Name_____

Lesson 9
Exercise 2

Missing Addition Number

Write the missing addition number in the box. The first one is done for you.

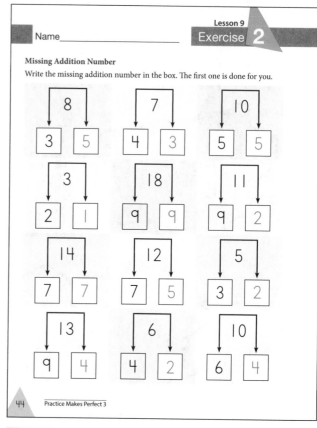

Name_____

Lesson 9
Exercise 3

Missing Addition Number

Write the missing addition number in the box.

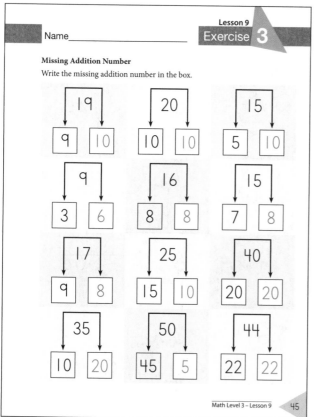

Name_____

Lesson 9
Exercise 4

Rounding to the nearest 100

Show your teacher how to round these numbers to the nearest 100. Circle the correct answer.

366	300 or (400)
931	(900) or 1,000
575	500 or (600)
183	100 or (200)
338	(300) or 400
274	200 or (300)

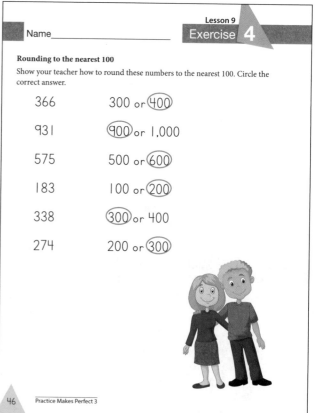

Solutions Manual: Lesson 10

Adding and Subtracting Larger Amounts of Money

Remember, these are the steps we follow when we add amounts of money.

Step 1: Line up the decimal point in the answer with the decimal points in the addends (the numbers being added together).

Step 2: Add from right to left, just like any other addition problem.

Step 3: Write the dollar sign in the answer (the sum).

Add the Money.

$$\begin{array}{r} {}^{1}\$36.23 \\ + \$\ 4.71 \\ \hline \$40.94 \end{array} \qquad \begin{array}{r} {}^{1\ 1}\$5.89 \\ + \$1.34 \\ \hline \$7.23 \end{array}$$

$$\begin{array}{r} {}^{1}\$22.13 \\ + \$18.45 \\ \hline \$40.58 \end{array} \qquad \begin{array}{r} \$91.03 \\ + \$42.91 \\ \hline \$133.94 \end{array}$$

Create your own money addition problems and solve them.

```
   $     .            $     .
+  $     .         +  $     .
```

Answers will vary.

When we subtract large amounts of money, we follow these steps:

Step 1: Line up the decimal point in the answer with the decimal points in the numbers that you are subtracting.

Step 2: Subtract from right to left.

Step 3: Write the dollar sign in the answer.

   Note: if you need to borrow, the same rules apply to subtracting money as they do in any other type of subtraction.

Subtract the Money.

$$\begin{array}{r} {}^{5}\$4\overset{}{6}.09 \\ - \$31.27 \\ \hline \$14.82 \end{array} \qquad \begin{array}{r} {}^{7}\$8\overset{}{6}.91 \\ - \$69.01 \\ \hline \$17.90 \end{array} \qquad \begin{array}{r} {}^{6}\$7\overset{}{1}.88 \\ - \$19.88 \\ \hline \$52.00 \end{array}$$

Rounding

Round these numbers to the nearest 100. Circle the correct answer.

456	400 or ⟨500⟩
821	⟨800⟩ or 900
699	600 or ⟨700⟩
101	⟨100⟩ or 200

Rounding and Mental Math Practice

Try to solve these in your brain.

Round these to the nearest 10. Tell your teacher the answers.

13 __10__   46 __50__

81 __80__   97 __100__

Round these to the nearest 100. Tell your teacher the answers.

846 __800__   912 __900__

277 __300__   538 __500__

High and Low

Write the numbers from least to greatest.

4,217   4,172   4,127   4,271

4,127	4,172	4,217	4,271
least	between	between	greatest

More or Less? – Number Sense

Do the addition and subtraction to fill in this chart. The first one is done for you.

	10 More	10 Less		10 More	10 Less
65	75	55	29	39	19
12	22	2	19	29	9
88	98	78	63	73	53
52	62	42	13	23	3
71	81	61	85	95	75
47	57	37	79	89	69
34	44	24	90	100	80

Solutions Manual: Lesson 11

Review of All New Concepts

Solve these addition problems. Show your work if you need to carry over.

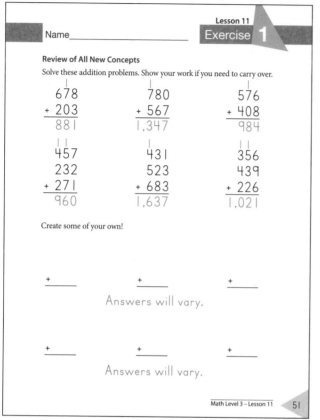

$$678 + 203 = 881$$

$$780 + 567 = 1,347$$

$$576 + 408 = 984$$

$$457 + 232 + 271 = 960$$

$$431 + 523 + 683 = 1,637$$

$$356 + 439 + 226 = 1,021$$

Create some of your own!

+ _____ + _____ + _____

Answers will vary.

+ _____ + _____ + _____

Answers will vary.

Solve these subtraction problems.

$$67 - 58 = 9$$

$$381 - 272 = 109$$

$$805 - 123 = 682$$

$$208 - 124 = 84$$

$$834 - 54 = 780$$

$$673 - 456 = 217$$

Create some of your own!

– _____ – _____ – _____

Answers will vary.

– _____ – _____ – _____

Answers will vary.

Figure it out!

Solve and round the answer to the nearest 100.

$$633 + 258 = 891$$ Rounds to 900

$$387 + 435 = 822$$ Rounds to 800

$$589 - 143 = 446$$ Rounds to 400

$$348 - 139 = 209$$ Rounds to 200

Rounding and In Between

Write the 10s the number is between, then write the ten the number is closer to.

46 is between ___40___ and ___50___ and closer to ___50___

74 is between ___70___ and ___80___ and closer to ___70___

83 is between ___80___ and ___90___ and closer to ___80___

Write the 100s the number is between, then write the hundred the number is closer to.

432 is between ___400___ and ___500___ and closer to ___400___

567 is between ___500___ and ___600___ and closer to ___600___

258 is between ___200___ and ___300___ and closer to ___300___

Solutions Manual: Lessons 12 – 13

Name_____

Introducing Multiplication of 0, 1, 2, and 5
Practice with addition and multiplication.

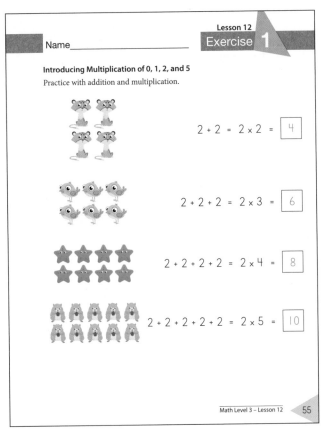

2 + 2 = 2 x 2 = 4

2 + 2 + 2 = 2 x 3 = 6

2 + 2 + 2 + 2 = 2 x 4 = 8

2 + 2 + 2 + 2 + 2 = 2 x 5 = 10

Math Level 3 – Lesson 12   55

Name_____

Matching
Draw a line from the multiplication problem to the addition problem it is equal to.

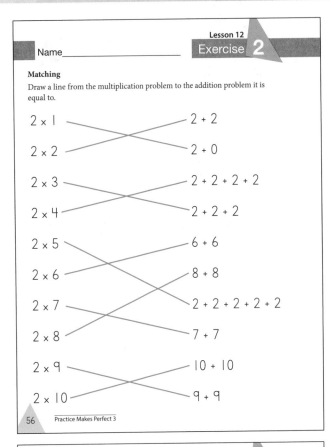

2 x 1   2 + 2
2 x 2   2 + 0
2 x 3   2 + 2 + 2 + 2
2 x 4   2 + 2 + 2
2 x 5   6 + 6
2 x 6   8 + 8
2 x 7   2 + 2 + 2 + 2 + 2
2 x 8   7 + 7
2 x 9   10 + 10
2 x 10   9 + 9

56   Practice Makes Perfect 3

Name_____

Multiplication Color By Number
Solve these multiplication problems to determine what colors to use!

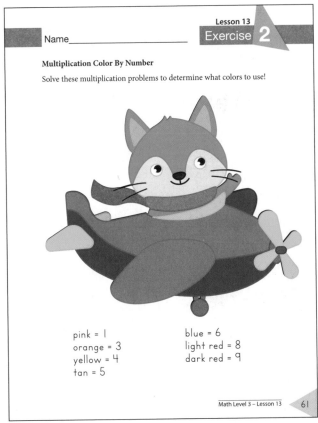

pink = 1   blue = 6
orange = 3   light red = 8
yellow = 4   dark red = 9
tan = 5

Math Level 3 – Lesson 13   61

Name_____

Fill in the Numbers
Use skip counting to fill in the missing numbers. Be sure to follow the arrows as you go.

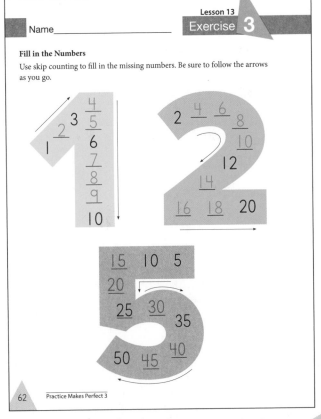

62   Practice Makes Perfect 3

Worksheet Solutions Manual   161

Solutions Manual: Lessons 13 – 14

Name_____

Lesson 13
Exercise **4**

Choose the sign
Should you add, subtract, multiply, or divide? Circle the correct sign for each problem.

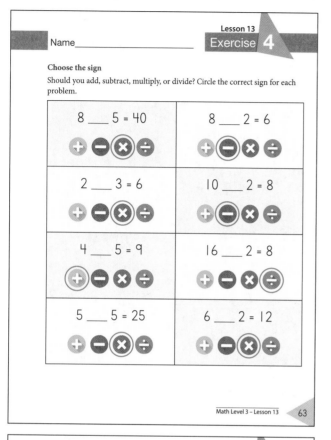

8 ___ 5 = 40 ⊕ ⊖ ⊗ ÷ (× circled)

8 ___ 2 = 6 ⊕ ⊖ ⊗ ÷ (− circled)

2 ___ 3 = 6 ⊕ ⊖ ⊗ ÷ (× circled)

10 ___ 2 = 8 ⊕ ⊖ ⊗ ÷ (− circled)

4 ___ 5 = 9 ⊕ ⊖ ⊗ ÷ (+ circled)

16 ___ 2 = 8 ⊕ ⊖ ⊗ ÷ (÷ circled)

5 ___ 5 = 25 ⊕ ⊖ ⊗ ÷ (× circled)

6 ___ 2 = 12 ⊕ ⊖ ⊗ ÷ (× circled)

Math Level 3 – Lesson 13   63

Name_____

Lesson 13
Exercise **5**

Multiplication Fact Practice
Write your answer to the following multiplication problems.

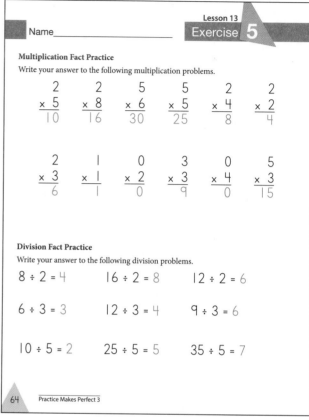

$\begin{array}{r} 2 \\ \times 5 \\ \hline 10 \end{array}$
$\begin{array}{r} 2 \\ \times 8 \\ \hline 16 \end{array}$
$\begin{array}{r} 5 \\ \times 6 \\ \hline 30 \end{array}$
$\begin{array}{r} 5 \\ \times 5 \\ \hline 25 \end{array}$
$\begin{array}{r} 2 \\ \times 4 \\ \hline 8 \end{array}$
$\begin{array}{r} 2 \\ \times 2 \\ \hline 4 \end{array}$

$\begin{array}{r} 2 \\ \times 3 \\ \hline 6 \end{array}$
$\begin{array}{r} 1 \\ \times 1 \\ \hline 1 \end{array}$
$\begin{array}{r} 0 \\ \times 2 \\ \hline 0 \end{array}$
$\begin{array}{r} 3 \\ \times 3 \\ \hline 9 \end{array}$
$\begin{array}{r} 0 \\ \times 4 \\ \hline 0 \end{array}$
$\begin{array}{r} 5 \\ \times 3 \\ \hline 15 \end{array}$

Division Fact Practice
Write your answer to the following division problems.

$8 \div 2 = 4$ $16 \div 2 = 8$ $12 \div 2 = 6$

$6 \div 3 = 3$ $12 \div 3 = 4$ $9 \div 3 = 6$

$10 \div 5 = 2$ $25 \div 5 = 5$ $35 \div 5 = 7$

64   Practice Makes Perfect 3

Name_____

Lesson 14
Exercise **1**

Introducing Multiplying and Dividing by 10
Counting by 10s.

10 _20_ _30_ _40_ _50_ _60_ _70_ _80_ _90_ 100
(10×1) (10×2) (10×3) (10×4) (10×5) (10×6) (10×7) (10×8) (10×9) (10×10)

Multiply by 10.

$5 \times 10 = \underline{50}$ $6 \times 10 = \underline{60}$ $4 \times 10 = \underline{40}$

$7 \times 10 = \underline{70}$ $8 \times 10 = \underline{80}$ $9 \times 10 = \underline{90}$

Add and Subtract. Watch your signs!

$\begin{array}{r} \overset{1\,1}{385} \\ 24 \\ + 19 \\ \hline 428 \end{array}$
$\begin{array}{r} 8\overset{7}{2}9 \\ - 578 \\ \hline 251 \end{array}$
$\begin{array}{r} 516 \\ + 322 \\ \hline 838 \end{array}$
$\begin{array}{r} \overset{8\,9}{9}00 \\ - 333 \\ \hline 567 \end{array}$

Rounding
Round these to the nearest 10.

22 _20_   91 _90_

Round these to the nearest 100.

240 _200_   852 _900_

Math Level 3 – Lesson 14   65

Name_____

Lesson 14
Exercise **2**

How Many and How Much? Solve the following problems. Watch the signs! The first one is done for you.

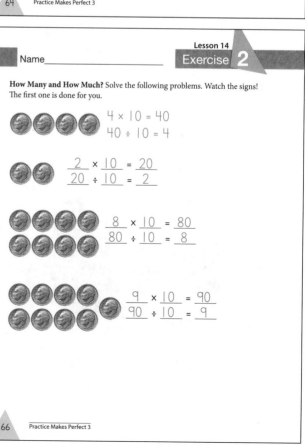

$4 \times 10 = 40$
$40 \div 10 = 4$

$\underline{2} \times \underline{10} = \underline{20}$
$\underline{20} \div \underline{10} = \underline{2}$

$\underline{8} \times \underline{10} = \underline{80}$
$\underline{80} \div \underline{10} = \underline{8}$

$\underline{9} \times \underline{10} = \underline{90}$
$\underline{90} \div \underline{10} = \underline{9}$

66   Practice Makes Perfect 3

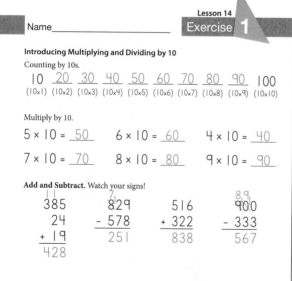

Solutions Manual: Lessons 14 –15

Name_____

Mutt Math: Multiplication
Today, let's practice multiplication facts. Multiply the numbers on the outside of the wheel by the number in the center. Put your answers in the outer circle.

Name_____

Mutt Math: Division
Today, let's practice division facts. Divide the numbers on the outside of the wheel by the number in the center. Put your answers in the outer circle.

Name_____

Math Crossword. Fill in the missing squares with the correct number.

Multiplication and Division. Solve these multiplication problems. Underneath each one, write a matching division problem.

2 x 3 = 6   2 x 5 = 10   2 x 7 = 14

6 ÷ 2 = 3   10 ÷ 2 = 5   14 ÷ 2 = 7

Name_____

Mental Math Mania!
Solve each mental math number sentence. Work from left to right. Each time you get one right, do two jumping jacks. Tell your teacher the correct answer.

$10 \times 2 - 5 + 10 - 4 = 21$

$25 \div 5 + 10 + 5 + 10 \div 3 = 10$

$18 - 9 \times 10 + 5 + 3 + 2 = 100$

Area and Perimeter
Figure out the area of a square that has 5-foot sides.

5 x 5 = 25 feet

Now, decide the perimeter of a rectangle that has a 10-yard side and an 8-yard side.

10 + 10 = 20
8 + 8 = 16

20 + 16 = 36 yards

Solutions Manual: Lesson 16

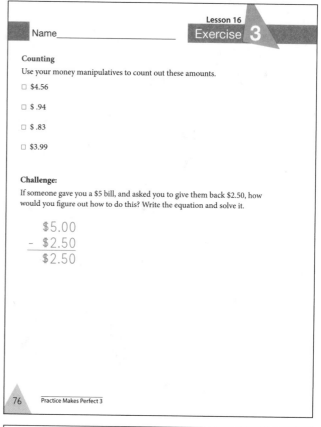

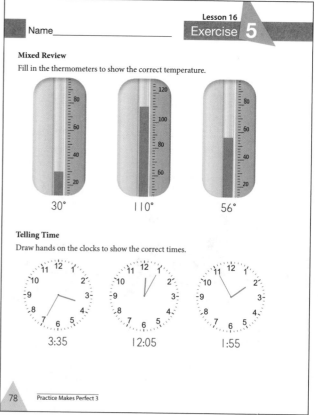

Name_____   Lesson 16   Exercise **3**

Rounding. Round these to the nearest 10.

32 __30__   76 __80__

91 __90__   14 __10__

29 __30__   82 __80__

Round these to the nearest 100.

423 __400__   755 __800__

910 __900__   132 __100__

377 __400__   821 __800__

How Long. Solve this story problem.

Hairo volunteered to sweep three of the hallways in the children's home. He started his chore at 1:20 in the afternoon and finished at 3:45. Draw hands on the first clock to show the time when he started. Draw hands on the second clock to show when he finished. How long did it take him to sweep the 3 hallways?

2 hours
and
25 minutes

Name_____   Lesson 16   Exercise **3**

Counting

Use your money manipulatives to count out these amounts.

☐ $4.56

☐ $ .94

☐ $ .83

☐ $3.99

Challenge:

If someone gave you a $5 bill, and asked you to give them back $2.50, how would you figure out how to do this? Write the equation and solve it.

$$\begin{array}{r} \$5.00 \\ - \ \$2.50 \\ \hline \$2.50 \end{array}$$

Name_____   Lesson 16   Exercise **4**

Mental Math

Solve these number sentences from left to right. Be sure to tell your teacher the correct answer.

30 ÷ 3 + 10 + 4 ÷ 3 = 8

10 + 4 ÷ 2 = 7

12 − 6 + 4 + 5 = 15

Add and Subtract. Watch the signs. Circle the answers that are odd with a green pencil and the even answers with a red pencil.

$$\begin{array}{r} \overset{2\ 13}{3\cancel{4}2} \\ - \ 256 \\ \hline \boxed{86} \end{array} \qquad \begin{array}{r} \overset{6\ 1\ 1}{7\cancel{2}3} \\ - \ 545 \\ \hline \boxed{178} \end{array} \qquad \begin{array}{r} 1\ 1 \\ 834 \\ 75 \\ + \ 34 \\ \hline \boxed{943} \end{array}$$

$$\begin{array}{r} 566 \\ - \ 423 \\ \hline \boxed{143} \end{array} \qquad \begin{array}{r} 622 \\ + \ 872 \\ \hline \boxed{1,494} \end{array} \qquad \begin{array}{r} 1\ 1 \\ 423 \\ 123 \\ + \ 164 \\ \hline \boxed{710} \end{array}$$

Name_____   Lesson 16   Exercise **5**

Mixed Review

Fill in the thermometers to show the correct temperature.

30°   110°   56°

Telling Time

Draw hands on the clocks to show the correct times.

3:35   12:05   1:55

Solutions Manual: Lesson 17

Lesson 17
Exercise **1**

Name_____

Taking Fractions Deeper

In this lesson, you will being doing some in-depth exploration of fractional concepts. You may enjoy completing a section of these reviews each day after you work with the fractional concepts in your *Math Lessons* book.

Counting. Count the amount of money in the piggy banks.

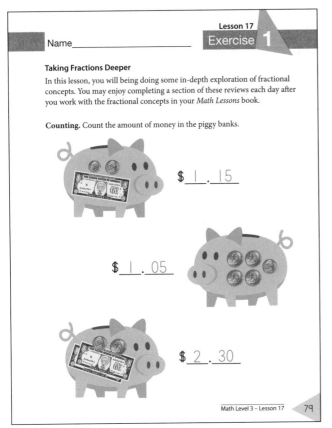

$ 1 . 15

$ 1 . 05

$ 2 . 30

Math Level 3 – Lesson 17   79

Lesson 17
Exercise **2**

Name_____

Time Concept Practice

Write down the correct time of each clock.

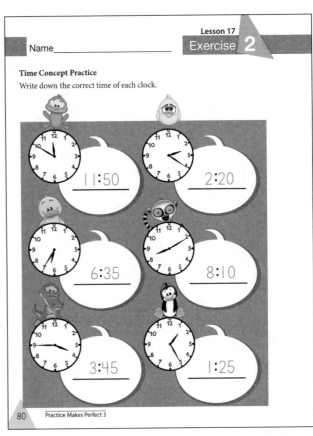

11:50

2:20

6:35

8:10

3:45

1:25

80   Practice Makes Perfect 3

Lesson 17
Exercise **3**

Name_____

Ordering Numbers

Place the numbers in the correct order from smallest to largest.

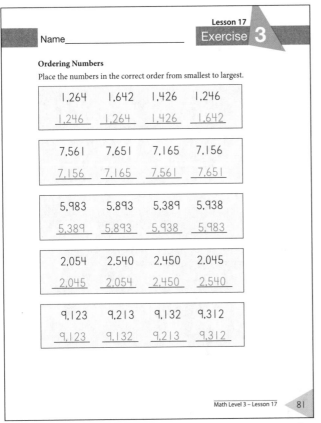

1,264	1,642	1,426	1,246
1,246	1,264	1,426	1,642

7,561	7,651	7,165	7,156
7,156	7,165	7,561	7,651

5,983	5,893	5,389	5,938
5,389	5,893	5,938	5,983

2,054	2,540	2,450	2,045
2,045	2,054	2,450	2,540

9,123	9,213	9,132	9,312
9,123	9,132	9,213	9,312

Math Level 3 – Lesson 17   81

Lesson 17
Exercise **4**

Name_____

Place Value

Use your Place Value Slider to show your teacher these numbers. Explain what each digit of each number stands for.

4,523   8,402   1,253   956

Practice with Fractional Concepts

Circle groups of 10 triangles and answer the questions.

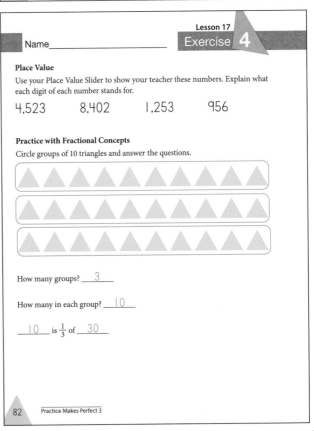

How many groups? ___3___

How many in each group? ___10___

___10___ is $\frac{1}{3}$ of ___30___

82   Practice Makes Perfect 3

Solutions Manual: Lessons 18 – 19

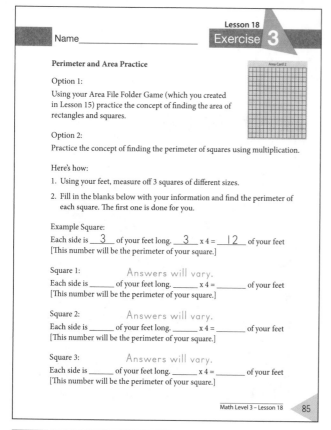

Perimeter and Area Practice

Option 1:

Using your Area File Folder Game (which you created in Lesson 15) practice the concept of finding the area of rectangles and squares.

Area Card 2

Option 2:

Practice the concept of finding the perimeter of squares using multiplication.

Here's how:

1. Using your feet, measure off 3 squares of different sizes.

2. Fill in the blanks below with your information and find the perimeter of each square. The first one is done for you.

Example Square:

Each side is __3__ of your feet long. __3__ x 4 = __12__ of your feet [This number will be the perimeter of your square.]

Square 1:   *Answers will vary.*
Each side is _____ of your feet long. _____ x 4 = _____ of your feet [This number will be the perimeter of your square.]

Square 2:   *Answers will vary.*
Each side is _____ of your feet long. _____ x 4 = _____ of your feet [This number will be the perimeter of your square.]

Square 3:   *Answers will vary.*
Each side is _____ of your feet long. _____ x 4 = _____ of your feet [This number will be the perimeter of your square.]

Name the Sign

Write the missing signs.

$36 \div 9 = 4$ $10 \times 4 = 40$ $16 \div 4 = 4$

$10 - 4 = 6$ $8 \div 2 = 4$ $4 + 6 = 10$

Fill in the Blanks

Multiplying and dividing by 4.

$4 \times 1 = \underline{4}$ $12 \div 4 = \underline{3}$

$4 \times 3 = \underline{12}$ $8 \div 4 = \underline{2}$

$4 \times 6 = \underline{24}$ $16 \div 4 = \underline{4}$

$4 \times 9 = \underline{36}$ $4 \div 4 = \underline{1}$

$4 \times 4 = \underline{16}$ $20 \div 4 = \underline{5}$

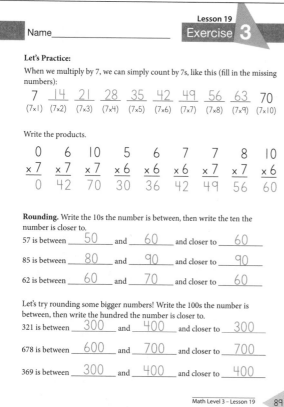

Let's Practice:

When we multiply by 7, we can simply count by 7s, like this (fill in the missing numbers):

7   14   21   28   35   42   49   56   63   70
(7×1) (7×2) (7×3) (7×4) (7×5) (7×6) (7×7) (7×8) (7×9) (7×10)

Write the products.

| $\frac{\times 7}{0}$ 0 | $\frac{\times 7}{42}$ 6 | $\frac{\times 7}{70}$ 10 | $\frac{\times 6}{30}$ 5 | $\frac{\times 6}{36}$ 6 | $\frac{\times 6}{42}$ 7 | $\frac{\times 7}{49}$ 7 | $\frac{\times 7}{56}$ 8 | $\frac{\times 6}{60}$ 10 |

Rounding. Write the 10s the number is between, then write the ten the number is closer to.

57 is between __50__ and __60__ and closer to __60__

85 is between __80__ and __90__ and closer to __90__

62 is between __60__ and __70__ and closer to __60__

Let's try rounding some bigger numbers! Write the 100s the number is between, then write the hundred the number is closer to.

321 is between __300__ and __400__ and closer to __300__

678 is between __600__ and __700__ and closer to __700__

369 is between __300__ and __400__ and closer to __400__

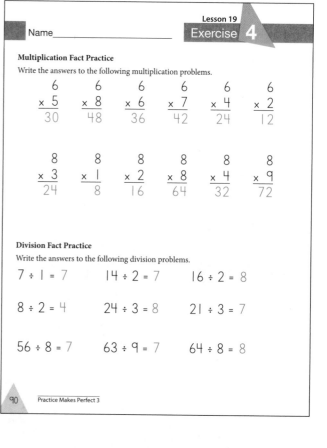

Multiplication Fact Practice

Write the answers to the following multiplication problems.

6	6	6	6	6	6
× 5	× 8	× 6	× 7	× 4	× 2
30	48	36	42	24	12

8	8	8	8	8	8
× 3	× 1	× 2	× 8	× 4	× 9
24	8	16	64	32	72

Division Fact Practice

Write the answers to the following division problems.

$7 \div 1 = 7$ $14 \div 2 = 7$ $16 \div 2 = 8$

$8 \div 2 = 4$ $24 \div 3 = 8$ $21 \div 3 = 7$

$56 \div 8 = 7$ $63 \div 9 = 7$ $64 \div 8 = 8$

Solutions Manual: Lessons 20 – 21

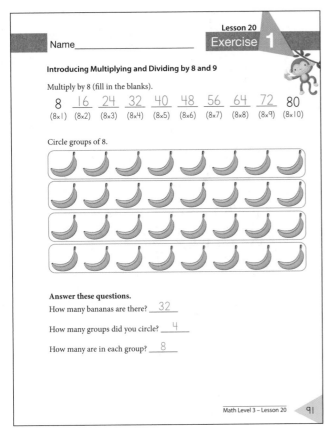

Introducing Multiplying and Dividing by 8 and 9

Multiply by 8 (fill in the blanks).

8   16   24   32   40   48   56   64   72   80
(8×1) (8×2) (8×3) (8×4) (8×5) (8×6) (8×7) (8×8) (8×9) (8×10)

Circle groups of 8.

Answer these questions.

How many bananas are there? __32__

How many groups did you circle? __4__

How many are in each group? __8__

Math Level 3 – Lesson 20   91

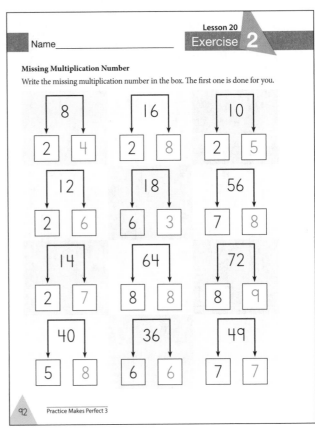

Missing Multiplication Number

Write the missing multiplication number in the box. The first one is done for you.

8	16	10
2 4	2 8	2 5
12	18	56
2 6	6 3	7 8
14	64	72
2 7	8 8	8 9
40	36	49
5 8	6 6	7 7

92   Practice Makes Perfect 3

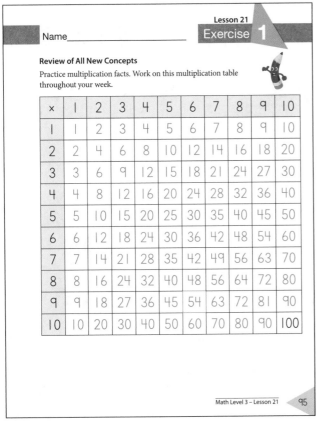

Review of All New Concepts

Practice multiplication facts. Work on this multiplication table throughout your week.

×	1	2	3	4	5	6	7	8	9	10
1	1	2	3	4	5	6	7	8	9	10
2	2	4	6	8	10	12	14	16	18	20
3	3	6	9	12	15	18	21	24	27	30
4	4	8	12	16	20	24	28	32	36	40
5	5	10	15	20	25	30	35	40	45	50
6	6	12	18	24	30	36	42	48	54	60
7	7	14	21	28	35	42	49	56	63	70
8	8	16	24	32	40	48	56	64	72	80
9	9	18	27	36	45	54	63	72	81	90
10	10	20	30	40	50	60	70	80	90	100

Math Level 3 – Lesson 21   95

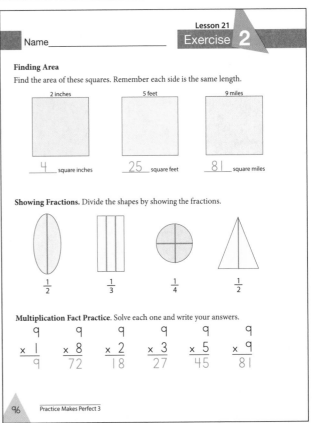

Finding Area

Find the area of these squares. Remember each side is the same length.

2 inches   5 feet   9 miles

__4__ square inches   __25__ square feet   __81__ square miles

Showing Fractions. Divide the shapes by showing the fractions.

$\frac{1}{2}$ $\frac{1}{3}$ $\frac{1}{4}$ $\frac{1}{2}$

Multiplication Fact Practice. Solve each one and write your answers.

$\begin{array}{r} 9 \\ \times\ 1 \\ \hline 9 \end{array}$
$\begin{array}{r} 9 \\ \times\ 8 \\ \hline 72 \end{array}$
$\begin{array}{r} 9 \\ \times\ 2 \\ \hline 18 \end{array}$
$\begin{array}{r} 9 \\ \times\ 3 \\ \hline 27 \end{array}$
$\begin{array}{r} 9 \\ \times\ 5 \\ \hline 45 \end{array}$
$\begin{array}{r} 9 \\ \times\ 9 \\ \hline 81 \end{array}$

96   Practice Makes Perfect 3

Solutions Manual: Lessons 21 – 22

Name_____

Counting Money
Count the amount of money in the piggy banks.

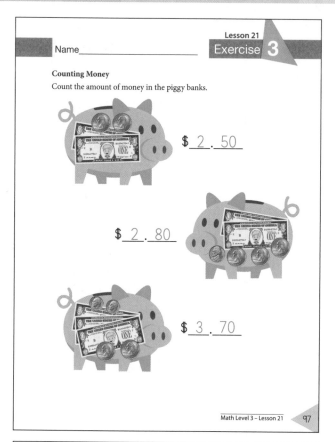

$ _2_._50_

$ _2_._80_

$ _3_._70_

Name_____

Understanding a graph. The children helped Dad to paint the rooms at the clinic. Study the graph below and answer the questions. Be sure to use the color below the paint bucket to determine your answer.

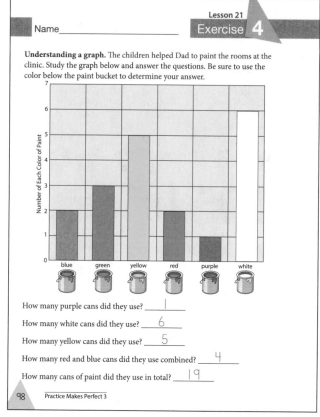

How many purple cans did they use? ___1___

How many white cans did they use? ___6___

How many yellow cans did they use? ___5___

How many red and blue cans did they use combined? ___4___

How many cans of paint did they use in total? ___19___

Name_____

Introducing Rounding to 1000s and Estimation
Place Value Review
Use your Place Value Slider to show your teacher these numbers.

5,612   7,810   9,672   3,991

Rounding
Round these to the nearest 10.

28 __30__   43 __40__

99 __100__   51 __50__

Challenge! 167 __170__

Round these to the nearest 100.

315 __300__   873 __900__

923 __900__   231 __200__

Challenge! 3,410 __3,400__

Name_____

Critical Thinking
When the storm knocked the electricity out, Mom helped Dad light 49 candles. Charlie, Hairo, and Charlotte worked together to set up 52 cots. Estimate how many candles and cots there were at the children's home.

Round (to the nearest 10) the number of candles. ___50___

Round (to the nearest 10) the number of cots. ___50___

Estimate how many there were all together. ___100___ (cots and candles)

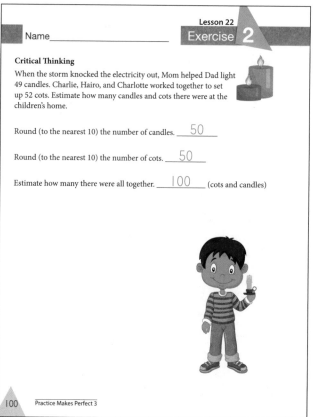

Solutions Manual: Lessons 22 – 23

Name_____

Missing dots. Fill in the missing dots to make the addition equation true.

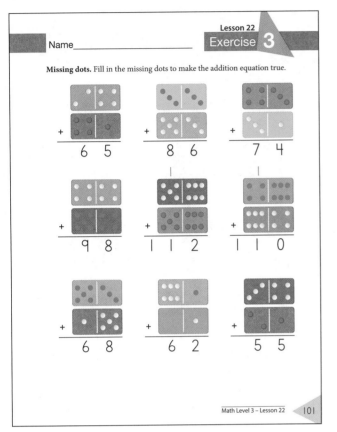

Name_____

Introducing Higher Place Value through Millions
Multiplication Match-up
Draw a line from the multiplication problem to the addition problem it is equal to.

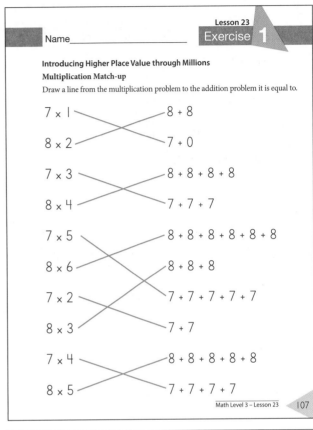

Name_____

Mutt Math: Division
Divide the numbers on the outside of the wheel by the number in the center. Write your answer in the outer ring. The first one is done for you.

Name_____

Place Value Practice and Exploration

278 = __2__ hundreds __7__ tens __8__ ones

1,629 = __1__ thousands __6__ hundreds __2__ tens __9__ ones

70 = __7__ tens __0__ ones

7 hundreds 5 tens 8 ones = ___758___

4 tens 3 ones = ___43___

100 + 20 + 5 = ___125___

400 + 10 + 3 = ___413___

Use your Place Value Slider to show your teacher these numbers.

3,419   9,002   5,555   7,126

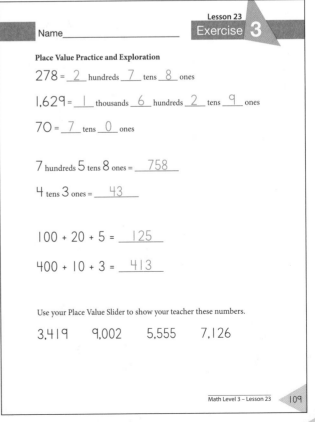

Solutions Manual: Lessons 23 – 24

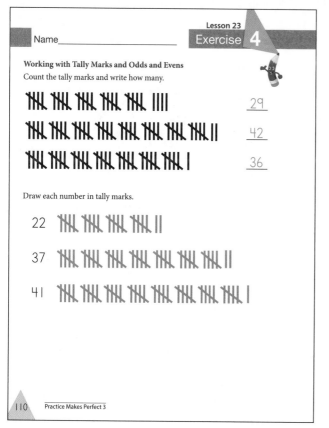

Working with Tally Marks and Odds and Evens
Count the tally marks and write how many.

卌 卌 卌 卌 卌 IIII — 29

卌 卌 卌 卌 卌 卌 卌 卌 II — 42

卌 卌 卌 卌 卌 卌 卌 I — 36

Draw each number in tally marks.

22 卌 卌 卌 卌 II

37 卌 卌 卌 卌 卌 卌 卌 II

41 卌 卌 卌 卌 卌 卌 卌 卌 I

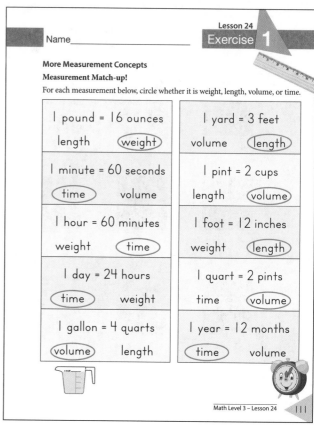

More Measurement Concepts
Measurement Match-up!
For each measurement below, circle whether it is weight, length, volume, or time.

1 pound = 16 ounces	1 yard = 3 feet
length (weight)	volume (length)
1 minute = 60 seconds	1 pint = 2 cups
(time) volume	length (volume)
1 hour = 60 minutes	1 foot = 12 inches
weight (time)	weight (length)
1 day = 24 hours	1 quart = 2 pints
(time) weight	time (volume)
1 gallon = 4 quarts	1 year = 12 months
(volume) length	(time) volume

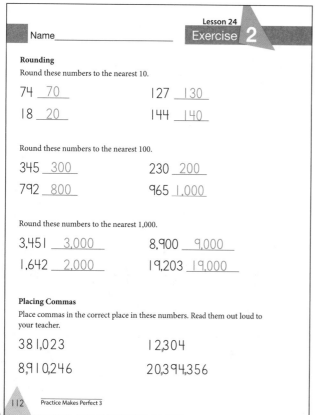

Rounding
Round these numbers to the nearest 10.

74 _70_   127 _130_
18 _20_   144 _140_

Round these numbers to the nearest 100.

345 _300_   230 _200_
792 _800_   965 _1,000_

Round these numbers to the nearest 1,000.

3,451 _3,000_   8,900 _9,000_
1,642 _2,000_   19,203 _19,000_

Placing Commas
Place commas in the correct place in these numbers. Read them out loud to your teacher.

381,023   12,304

8,910,246   20,394,356

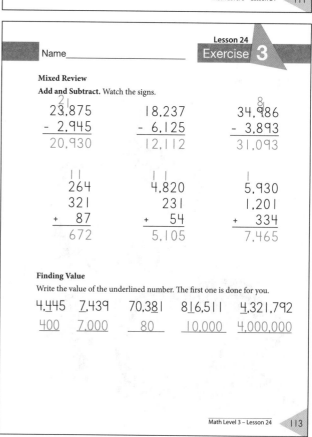

Mixed Review
Add and Subtract. Watch the signs.

```
  2 1
23,875        18,237        34,986
- 2,945       - 6,125       - 3,893
20,930        12,112        31,093
```

```
 1 1            1 1           1
 264           4,820         5,930
 321            231          1,201
+ 87          +  54         +  334
 672           5,105         7,465
```

Finding Value
Write the value of the underlined number. The first one is done for you.

4,4̲45   7̲,439   70,3̲81   8̲16,511   4̲,321,792
_400_   _7,000_   _80_   _10,000_   _4,000,000_

Solutions Manual: Lessons 24 – 25

Measurement Multiple Choice
Circle whether the object should be measured in inches, feet, yards, or miles.

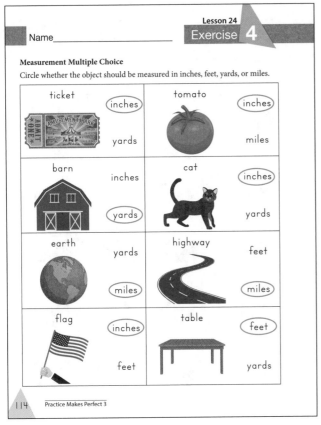

Introducing Solving for Unknowns
Fill in the missing number to make the equation true.

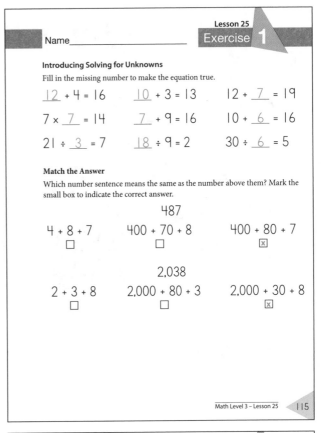

$\underline{12} + 4 = 16$ $\underline{10} + 3 = 13$ $12 + \underline{7} = 19$

$7 \times \underline{7} = 14$ $\underline{7} + 9 = 16$ $10 + \underline{6} = 16$

$21 \div \underline{3} = 7$ $\underline{18} \div 9 = 2$ $30 \div \underline{6} = 5$

Match the Answer
Which number sentence means the same as the number above them? Mark the small box to indicate the correct answer.

487

| 4 + 8 + 7 | 400 + 70 + 8 | 400 + 80 + 7 |
| □ | □ | ☒ |

2,038

| 2 + 3 + 8 | 2,000 + 80 + 3 | 2,000 + 30 + 8 |
| □ | □ | ☒ |

Adding and Subtracting. Complete your problems. If the answer is odd, underline it with a green pencil or crayon. If the answer is even, underline it with a red pencil.

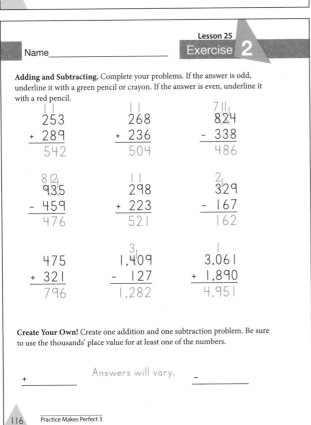

$\begin{array}{r} 253 \\ + 289 \\ \hline 542 \end{array}$ $\begin{array}{r} 268 \\ + 236 \\ \hline 504 \end{array}$ $\begin{array}{r} 824 \\ - 338 \\ \hline 486 \end{array}$

$\begin{array}{r} 935 \\ - 459 \\ \hline 476 \end{array}$ $\begin{array}{r} 298 \\ + 223 \\ \hline 521 \end{array}$ $\begin{array}{r} 329 \\ - 167 \\ \hline 162 \end{array}$

$\begin{array}{r} 475 \\ + 321 \\ \hline 796 \end{array}$ $\begin{array}{r} 1,409 \\ - 127 \\ \hline 1,282 \end{array}$ $\begin{array}{r} 3,061 \\ + 1,890 \\ \hline 4,951 \end{array}$

Create Your Own! Create one addition and one subtraction problem. Be sure to use the thousands' place value for at least one of the numbers.

_____ + _____   Answers will vary.   _____ - _____

Multiplication Grids
Shade each grid to match the multiplication fact and write in the missing answer. The first one is done for you.

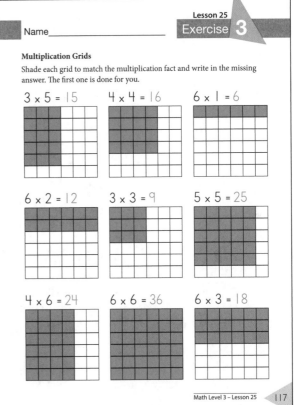

$3 \times 5 = 15$ $4 \times 4 = 16$ $6 \times 1 = 6$

$6 \times 2 = 12$ $3 \times 3 = 9$ $5 \times 5 = 25$

$4 \times 6 = 24$ $6 \times 6 = 36$ $6 \times 3 = 18$

Solutions Manual: Lessons 25 – 26

Name_____

Exercise 4

Solve!
Solve the division problems to find out the correct colors.

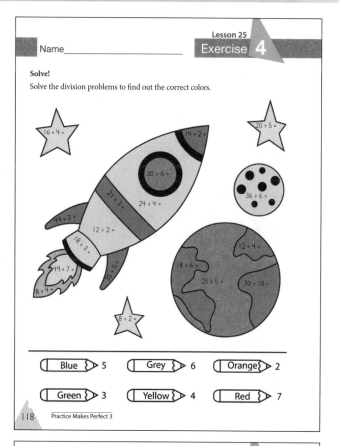

| Blue ▷ 5 | Grey ▷ 6 | Orange ▷ 2 |
| Green ▷ 3 | Yellow ▷ 4 | Red ▷ 7 |

Name_____

Exercise 1

Introducing Inequalities

Greater Than/Less Than. Circle the Greater Than Croc if the first number in each box is greater than the second. Circle the Less Than Croc if the first number is less. Remember, the wider side is the larger numbered side.

< = Less Than   Greater Than = >

6,832 ___ 6,382	2 × 4 ___ 3 × 3
7 × 7 ___ 6 × 8	25 ÷ 5 ___ 30 ÷ 3
1,903 ___ 1,390	64 ÷ 8 ___ 81 ÷ 9
10 ÷ 5 ___ 40 ÷ 8	3,812 ___ 3,481

Name_____

Exercise 2

Find the Sign!
In each box, circle the sign that makes the statement true.

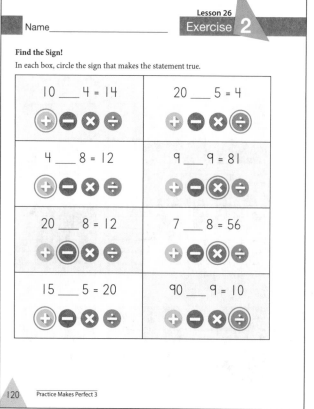

10 ___ 4 = 14	20 ___ 5 = 4
⊕ ⊖ ⊗ ÷	⊕ ⊖ ⊗ ÷
4 ___ 8 = 12	9 ___ 9 = 81
⊕ ⊖ ⊗ ÷	⊕ ⊖ ⊗ ÷
20 ___ 8 = 12	7 ___ 8 = 56
⊕ ⊖ ⊗ ÷	⊕ ⊖ ⊗ ÷
15 ___ 5 = 20	90 ___ 9 = 10
⊕ ⊖ ⊗ ÷	⊕ ⊖ ⊗ ÷

Name_____

Exercise 3

Multiplication with Missing Factors
Fill in the blank with the missing factor to make the equation true.

$6 × \underline{6} = 36$ $8 × \underline{8} = 64$

$3 × \underline{9} = 27$ $7 × \underline{9} = 63$

$8 × \underline{5} = 40$ $4 × \underline{5} = 20$

$9 × \underline{9} = 81$ $2 × \underline{12} = 24$

$7 × \underline{8} = 56$ $6 × \underline{6} = 36$

$2 × \underline{10} = 20$ $3 × \underline{10} = 30$

$5 × \underline{9} = 45$ $5 × \underline{6} = 30$

$4 × \underline{9} = 36$ $8 × \underline{9} = 72$

$3 × \underline{5} = 15$ $10 × \underline{10} = 100$

$9 × \underline{3} = 27$ $10 × \underline{6} = 60$

Solutions Manual: Lessons 26 – 27

Name_____

Find the Sign!

Charlotte loves to splash paint. She has painted equal and not equal signs. Circle the sign that she has painted to make the statement in each box true.

3 + 8 ___ 7 + 5	30 ÷ 10 ___ 9 ÷ 3
= ~~≠~~	= ~~≠~~
6 + 7 ___ 3 + 9	25 ÷ 5 ___ 50 ÷ 10
= ~~≠~~	= ~~≠~~

She has now painted greater than and less than signs. Circle the sign that she has painted to make the statement in each box true.

36 ÷ 4 ___ 3 × 5	20 ÷ 5 ___ 6 + 2
< ~~>~~	< ~~>~~
2 × 9 ___ 7 + 4	100 ÷ 10 ___ 4 × 2
< ~~>~~	< ~~>~~

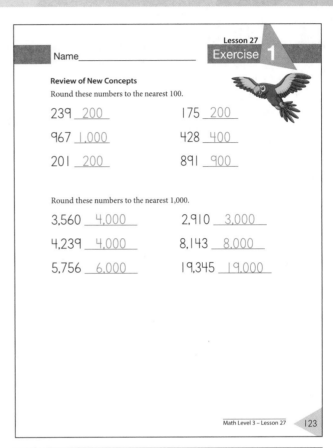

Name_____

Review of New Concepts

Round these numbers to the nearest 100.

239 _200_   175 _200_

967 _1,000_   428 _400_

201 _200_   891 _900_

Round these numbers to the nearest 1,000.

3,560 _4,000_   2,910 _3,000_

4,239 _4,000_   8,143 _8,000_

5,756 _6,000_   19,345 _19,000_

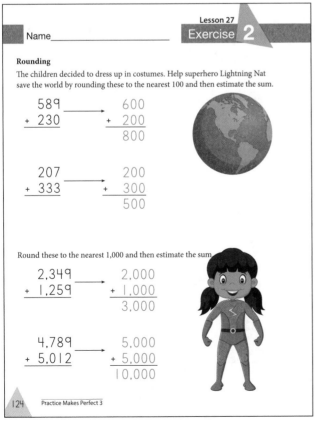

Name_____

Rounding

The children decided to dress up in costumes. Help superhero Lightning Nat save the world by rounding these to the nearest 100 and then estimate the sum.

```
  589  ____      600
+ 230          + 200
                 800
```

```
  207  ____      200
+ 333          + 300
                 500
```

Round these to the nearest 1,000 and then estimate the sum.

```
  2,349  ____    2,000
+ 1,259        + 1,000
                 3,000
```

```
  4,789  ____    5,000
+ 5,012        + 5,000
                10,000
```

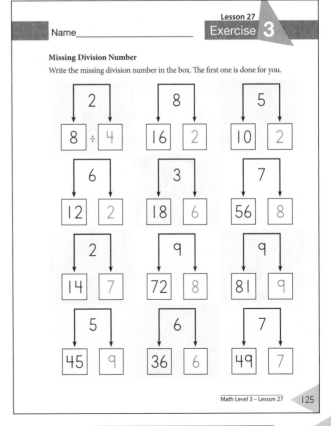

Name_____

Missing Division Number

Write the missing division number in the box. The first one is done for you.

| 2 | 8 | 5 |
| 8 ÷ 4 | 16   2 | 10   2 |

| 6 | 3 | 7 |
| 12   2 | 18   6 | 56   8 |

| 2 | 9 | 9 |
| 14   7 | 72   8 | 81   9 |

| 5 | 6 | 7 |
| 45   9 | 36   6 | 49   7 |

Solutions Manual: Lessons 27 – 28

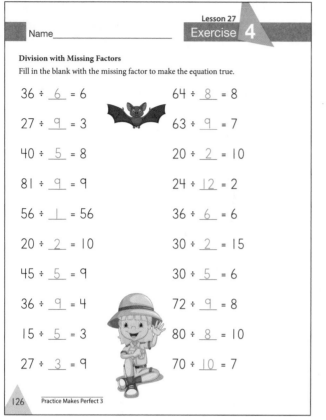

Name_____

Division with Missing Factors
Fill in the blank with the missing factor to make the equation true.

$36 \div \underline{6} = 6$ $64 \div \underline{8} = 8$

$27 \div \underline{9} = 3$ $63 \div \underline{9} = 7$

$40 \div \underline{5} = 8$ $20 \div \underline{2} = 10$

$81 \div \underline{9} = 9$ $24 \div \underline{12} = 2$

$56 \div \underline{1} = 56$ $36 \div \underline{6} = 6$

$20 \div \underline{2} = 10$ $30 \div \underline{2} = 15$

$45 \div \underline{5} = 9$ $30 \div \underline{5} = 6$

$36 \div \underline{9} = 4$ $72 \div \underline{9} = 8$

$15 \div \underline{5} = 3$ $80 \div \underline{8} = 10$

$27 \div \underline{3} = 9$ $70 \div \underline{10} = 7$

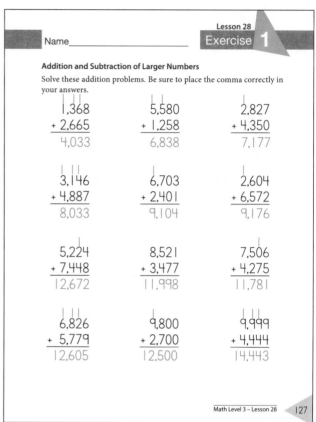

Name_____

Addition and Subtraction of Larger Numbers
Solve these addition problems. Be sure to place the comma correctly in your answers.

$$\begin{array}{r} 1,368 \\ + 2,665 \\ \hline 4,033 \end{array} \qquad \begin{array}{r} 5,580 \\ + 1,258 \\ \hline 6,838 \end{array} \qquad \begin{array}{r} 2,827 \\ + 4,350 \\ \hline 7,177 \end{array}$$

$$\begin{array}{r} 3,146 \\ + 4,887 \\ \hline 8,033 \end{array} \qquad \begin{array}{r} 6,703 \\ + 2,401 \\ \hline 9,104 \end{array} \qquad \begin{array}{r} 2,604 \\ + 6,572 \\ \hline 9,176 \end{array}$$

$$\begin{array}{r} 5,224 \\ + 7,448 \\ \hline 12,672 \end{array} \qquad \begin{array}{r} 8,521 \\ + 3,477 \\ \hline 11,998 \end{array} \qquad \begin{array}{r} 7,506 \\ + 4,275 \\ \hline 11,781 \end{array}$$

$$\begin{array}{r} 6,826 \\ + 5,779 \\ \hline 12,605 \end{array} \qquad \begin{array}{r} 9,800 \\ + 2,700 \\ \hline 12,500 \end{array} \qquad \begin{array}{r} 9,999 \\ + 4,444 \\ \hline 14,443 \end{array}$$

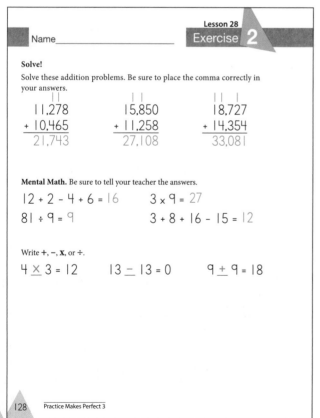

Name_____

Solve!
Solve these addition problems. Be sure to place the comma correctly in your answers.

$$\begin{array}{r} 11,278 \\ + 10,465 \\ \hline 21,743 \end{array} \qquad \begin{array}{r} 15,850 \\ + 11,258 \\ \hline 27,108 \end{array} \qquad \begin{array}{r} 18,727 \\ + 14,354 \\ \hline 33,081 \end{array}$$

Mental Math. Be sure to tell your teacher the answers.

$12 + 2 - 4 + 6 = 16$ $3 \times 9 = 27$

$81 \div 9 = 9$ $3 + 8 + 16 - 15 = 12$

Write **+, −, x,** or **÷**.

$4 \underline{\times} 3 = 12$ $13 \underline{-} 13 = 0$ $9 \underline{+} 9 = 18$

Name_____

Solve!
Solve these subtraction problems. Be sure to place the comma correctly in your answers.

$$\begin{array}{r} 6,345 \\ - 4,131 \\ \hline 2,214 \end{array} \qquad \begin{array}{r} 7,178 \\ - 5,045 \\ \hline 2,133 \end{array} \qquad \begin{array}{r} 6,842 \\ - 6,343 \\ \hline 499 \end{array}$$

$$\begin{array}{r} 7,234 \\ - 3,244 \\ \hline 3,990 \end{array} \qquad \begin{array}{r} 8,056 \\ - 6,158 \\ \hline 1,898 \end{array} \qquad \begin{array}{r} 3,344 \\ - 2,188 \\ \hline 1,156 \end{array}$$

$$\begin{array}{r} 2,467 \\ - 1,468 \\ \hline 999 \end{array} \qquad \begin{array}{r} 6,168 \\ - 3,269 \\ \hline 2,899 \end{array} \qquad \begin{array}{r} 8,765 \\ - 5,679 \\ \hline 3,086 \end{array}$$

$$\begin{array}{r} 5,734 \\ - 4,789 \\ \hline 945 \end{array} \qquad \begin{array}{r} 7,057 \\ - 4,371 \\ \hline 2,686 \end{array} \qquad \begin{array}{r} 9,873 \\ - 6,649 \\ \hline 3,224 \end{array}$$

Solutions Manual: Lessons 28 – 29

Clock and Time Concepts Practice

Fill in the blank with what the time will be and then, draw the hands in each clock showing the correct time.

Right now it is 3:15, in 45 minutes, it will be
4 : 00

Right now it is 7:30, in one hour and 10 minutes, it will be
8 : 40

Right now it is 12:50, in one hour and 20 minutes, it will be
2 : 10

Right now it is 6:55, in 50 minutes, it will be
7 : 45

Right now it is 9:00, in 2 hours and 15 minutes, it will be
11 : 15

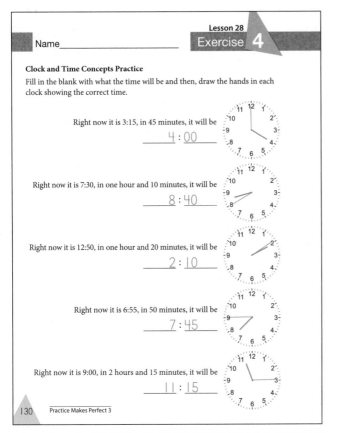

Introducing Roman Numerals

Match the Roman numerals with the standard numbers.

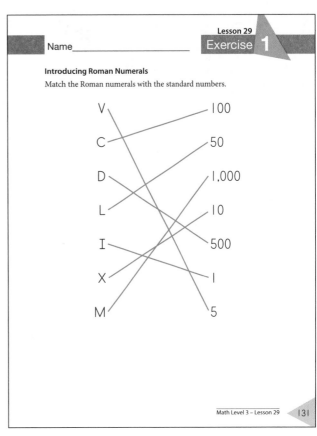

V	100
C	50
D	1,000
L	10
I	500
X	1
M	5

Multiplication

Practice multiplication facts. Fill in the chart with the correct numbers.

×	1	2	3	4	5	6	7	8	9	10
1	1	2	3	4	5	6	7	8	9	10
2	2	4	6	8	10	12	14	16	18	20
3	3	6	9	12	15	18	21	24	27	30
4	4	8	12	16	20	24	28	32	36	40
5	5	10	15	20	25	30	35	40	45	50

Roman Numerals

Convert these numbers into Roman numerals.

100 = C   50 = L
1,000 = M   500 = D

3 = III   12 = XII
7 = VII   21 = XXI
9 = IX   30 = XXX

What's the number? Fill in the blanks with the standard number.

C = 100   L = 50
M = 1,000   D = 500

IV = 4   VII = 7
IX = 9   III = 3
XX = 20   XVI = 16
LX = 60   XXI = 21

Multiplication

Practice multiplication facts. Fill in the chart with the correct numbers.

×	1	2	3	4	5	6	7	8	9	10
6	6	12	18	24	30	36	42	48	54	60
7	7	14	21	28	35	42	49	56	63	70
8	8	16	24	32	40	48	56	64	72	80
9	9	18	27	36	45	54	63	72	81	90
10	10	20	30	40	50	60	70	80	90	100

Solutions Manual: Lesson 30

Name_____

More about Roman Numerals

Solve the following problems. Be sure to check the signs to see if it is addition or subtraction.

```
   |               | |              | |
  5,320          7,102           8,201
  1,264          2,385           9,642
+ 4,800        + 5,173         + 6,381
 11,384         14,660          24,224


   | |             |              2 2
   933            821            667
   812            705            443
   755            623            559
 + 671          + 414          + 756
  3,171          2,563          2,425


  6 6 13          7              8 13
  X,X40          8,267          9,456
- 4,954        - 7,612        - 8,673
  2,786           655            783
```

Name_____

Money

Using play money, count out the following amounts.

$8.73 $ .82

$10.55 $5.19

Now work through these problems. Be sure to write your answers correctly with commas, dollar signs, or cent signs when needed.

If an item cost $2.20, and you paid with a $5 bill, what would be your change?

```
      4
  $5.00
- $2.20
  $2.80
```

If an item cost $9.62, and you paid with a $10 bill, what would be your change?

```
   9 9
 $10.00
- $9.62
  $0.38
```

If your total at the store was $62.38, and you paid with a $100 bill, what would be your change?

```
   9 9 9
 $100.00
- $62.38
  $37.62
```

Name_____

Color by Roman Numeral

Color the image below using Roman numerals to determine the colors.

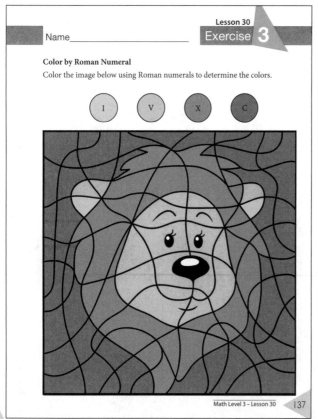

Name_____

Match the equation. Draw a line to match the equation or Roman numeral on the left with the matching number on the right.

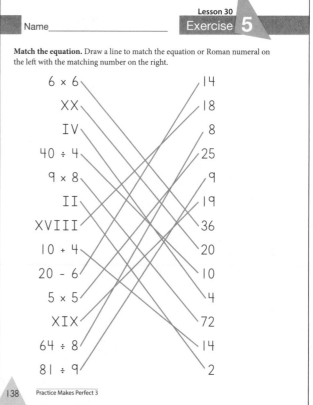

6×6 14
XX   18
IV   8
$40 \div 4$ 25
9×8 9
II   19
XVIII   36
$10 + 4$ 20
$20 - 6$ 10
5×5 4
XIX   72
$64 \div 8$ 14
$81 \div 9$ 2

Solutions Manual: Quizzes 1 – 2

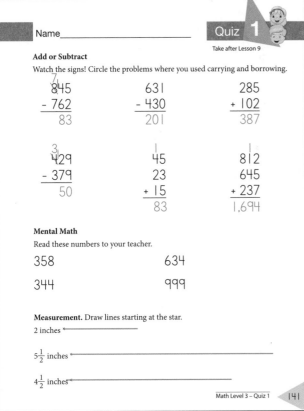

Add or Subtract

Watch the signs! Circle the problems where you used carrying and borrowing.

$$\begin{array}{r} {}^{7}8\!\!\!/45 \\ -\ 762 \\ \hline 83 \end{array} \qquad \begin{array}{r} 631 \\ -\ 430 \\ \hline 201 \end{array} \qquad \begin{array}{r} 285 \\ +\ 102 \\ \hline 387 \end{array}$$

$$\begin{array}{r} {}^{3}4\!\!\!/29 \\ -\ 379 \\ \hline 50 \end{array} \qquad \begin{array}{r} 45 \\ 23 \\ +\ 15 \\ \hline 83 \end{array} \qquad \begin{array}{r} 812 \\ 645 \\ +\ 237 \\ \hline 1,694 \end{array}$$

Mental Math

Read these numbers to your teacher.

358   634

344   999

Measurement. Draw lines starting at the star.

2 inches ✶———

$5\frac{1}{2}$ inches ✶————————————

$4\frac{1}{2}$ inches ✶——————————

Time

Fill in the clock faces to show the correct time.

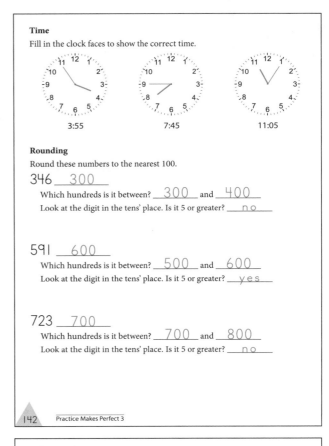

3:55   7:45   11:05

Rounding

Round these numbers to the nearest 100.

346 ___300___

   Which hundreds is it between? ___300___ and ___400___

   Look at the digit in the tens' place. Is it 5 or greater? ___no___

591 ___600___

   Which hundreds is it between? ___500___ and ___600___

   Look at the digit in the tens' place. Is it 5 or greater? ___yes___

723 ___700___

   Which hundreds is it between? ___700___ and ___800___

   Look at the digit in the tens' place. Is it 5 or greater? ___no___

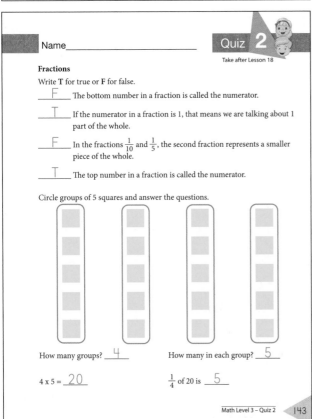

Fractions

Write **T** for true or **F** for false.

___F___ The bottom number in a fraction is called the numerator.

___T___ If the numerator in a fraction is 1, that means we are talking about 1 part of the whole.

___F___ In the fractions $\frac{1}{10}$ and $\frac{1}{5}$, the second fraction represents a smaller piece of the whole.

___T___ The top number in a fraction is called the numerator.

Circle groups of 5 squares and answer the questions.

How many groups? ___4___   How many in each group? ___5___

4 x 5 = ___20___ $\frac{1}{4}$ of 20 is ___5___

Perimeter and Area

If you had a square that had sides that were 3 feet long, what would be the perimeter? What would be the area?

Perimeter: ___3___ x ___4___ = ___12___ feet

Area: ___3___ x ___3___ = ___9___ square feet

Multiplication. Fill in the multiplication grid.

×	1	2	3	4	5	6	7	8	9	10
1	1	2	3	4	5	6	7	8	9	10
2	2	4	6	8	10	12	14	16	18	20
3	3	6	9	12	15	18	21	24	27	30
4	4	8	12	16	20	24	28	32	36	40
5	5	10	15	20	25	30	35	40	45	50
6	6	12	18	24	30	36	42	48	54	60
7	7	14	21	28	35	42	49	56	63	70
8	8	16	24	32	40	48	56	64	72	80
9	9	18	27	36	45	54	63	72	81	90
10	10	20	30	40	50	60	70	80	90	100

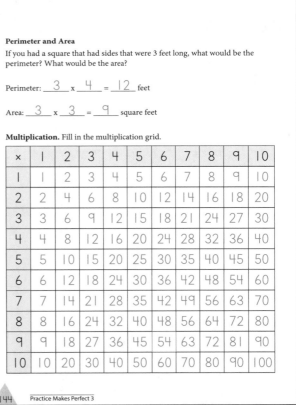

Solutions Manual: Quizzes 2 – 3

Divide

$30 \div 3 = 10$ $27 \div 3 = 9$

$15 \div 3 = 5$ $25 \div 5 = 5$

$20 \div 2 = 10$ $36 \div 4 = 9$

$12 \div 4 = 3$ $40 \div 10 = 4$

Critical Thinking

Show your teacher at least one multiplication fact to go with each of the division facts above. For example:

$3 \times 10 = 30$ $3 \times 9 = 27$

$10 \times 3 = 30$ $9 \times 3 = 27$

$3 \times 5 = 15$ $5 \times 5 = 25$

$5 \times 3 = 15$ $4 \times 9 = 36$

$2 \times 10 = 20$ $9 \times 4 = 36$

$10 \times 2 = 20$ $4 \times 10 = 40$

$3 \times 4 = 12$ $10 \times 4 = 40$

$4 \times 3 = 12$

Add or Subtract

$$\begin{array}{r} {}^{4}\cancel{5}{}^{15} \\ 56\cancel{2} \\ -\ 388 \\ \hline 174 \end{array} \qquad \begin{array}{r} 742 \\ -\ 540 \\ \hline 202 \end{array} \qquad \begin{array}{r} {}^{1}{}^{1} \\ 396 \\ +\ 204 \\ \hline 600 \end{array}$$

$$\begin{array}{r} 4{,}501 \\ +\ \ 320 \\ \hline 4{,}821 \end{array} \qquad \begin{array}{r} {}^{8} \\ 2{,}9\cancel{1}2 \\ -\ 1{,}890 \\ \hline 1{,}022 \end{array} \qquad \begin{array}{r} {}^{1} \\ 341 \\ 55 \\ +\ \ 23 \\ \hline 419 \end{array}$$

Take after Lesson 27

Place Value

Fill in the blanks.

35,120

In this number,
$3 = \underline{3}$ groups of $\underline{10{,}000}$.
$5 = \underline{5}$ groups of $\underline{1{,}000}$.
$1 = \underline{1}$ groups of $\underline{100}$.
$2 = \underline{2}$ groups of $\underline{10}$.
$0 = \underline{0}$ groups of $\underline{1}$.

5,623,890

In this number,
$5 = \underline{5}$ groups of $\underline{1{,}000{,}000}$.
$6 = \underline{6}$ groups of $\underline{100{,}000}$.
$2 = \underline{2}$ groups of $\underline{10{,}000}$.
$3 = \underline{3}$ groups of $\underline{1{,}000}$.
$8 = \underline{8}$ groups of $\underline{100}$.
$9 = \underline{9}$ groups of $\underline{10}$.
$0 = \underline{0}$ groups of $\underline{1}$.

What's the number? Read these numbers to your teacher.

32,678,103   10,592,682

768,012

What's the sign? Fill in > or <.

$\dfrac{1}{4} > \dfrac{1}{8}$ $12 \div 3 > 15 \div 5$

$18 < 81$

What's the sign? Fill in = or ≠.

$24 \div 6 \underline{=} 16 \div 4$

$64 \div 8 \underline{\neq} 27 \div 3$

$1 \text{ ton} \underline{=} 2{,}000 \text{ pounds}$

$1 \text{ mile} \underline{=} 5{,}280 \text{ feet}$

$1{,}760 \text{ yards} \underline{=} 1 \text{ mile}$

$7 \text{ feet} \underline{\neq} 2 \text{ yards}$

Solve. Solve these story problems.

Mom bought the children a treat after their walk around Lima's market. The treat cost $8.76. If she paid with a $10 bill, how much was her change?

$$\begin{array}{r} {}^{9} \\ \$1\cancel{0}.\cancel{0}\cancel{0} \\ -\ \$8.76 \\ \hline \$1.24 \end{array}$$

The local churches raised $450 to go towards building the clinic at the children's home. The electrician's work cost $389. How much was left over after paying the electrician for his work?

$$\begin{array}{r} {}^{3}\ {}^{14} \\ \$4\cancel{5}0.00 \\ -\ \$389.00 \\ \hline \$61.00 \end{array}$$

Solutions Manual: Quiz 4

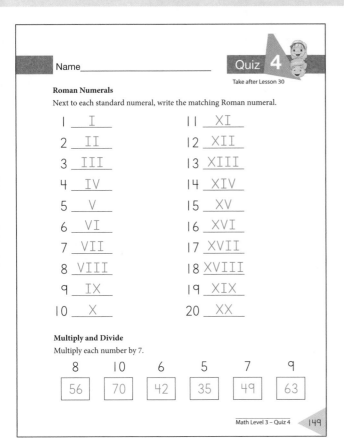

Name_____

Take after Lesson 30

Roman Numerals

Next to each standard numeral, write the matching Roman numeral.

1	_I_	11	_XI_
2	_II_	12	_XII_
3	_III_	13	_XIII_
4	_IV_	14	_XIV_
5	_V_	15	_XV_
6	_VI_	16	_XVI_
7	_VII_	17	_XVII_
8	_VIII_	18	_XVIII_
9	_IX_	19	_XIX_
10	_X_	20	_XX_

Multiply and Divide

Multiply each number by 7.

8	10	6	5	7	9
56	70	42	35	49	63

Math Level 3 – Quiz 4   149

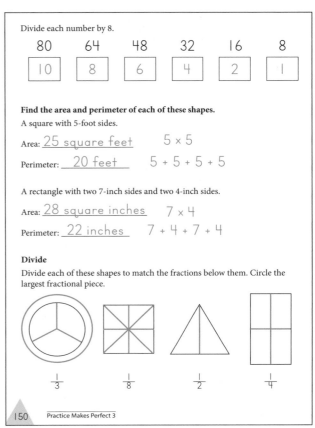

Divide each number by 8.

80	64	48	32	16	8
10	8	6	4	2	1

Find the area and perimeter of each of these shapes.

A square with 5-foot sides.

Area: _25 square feet_   5 × 5

Perimeter: _20 feet_   5 + 5 + 5 + 5

A rectangle with two 7-inch sides and two 4-inch sides.

Area: _28 square inches_   7 × 4

Perimeter: _22 inches_   7 + 4 + 7 + 4

Divide

Divide each of these shapes to match the fractions below them. Circle the largest fractional piece.

$\frac{1}{3}$ $\frac{1}{8}$ $\frac{1}{2}$ $\frac{1}{4}$

150   Practice Makes Perfect 3

Quiz Solutions Manual   179

LEVELS K-6
MATH LESSONS FOR A LIVING EDUCATION
A CHARLOTTE MASON FLAVOR TO MATH FOR TODAY'S STUDENT

Level K, Kindergarten
978-1-68344-176-2

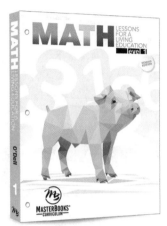

Level 1, Grade 1
978-0-89051-923-3

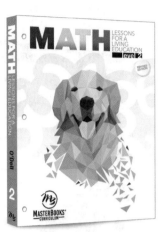

Level 2, Grade 2
978-0-89051-924-0

Level 3, Grade 3
978-0-89051-925-7

Level 4, Grade 4
978-0-89051-926-4

Level 5, Grade 5
978-0-89051-927-1

ATTRACTIVE FULL-COLOR LESSONS

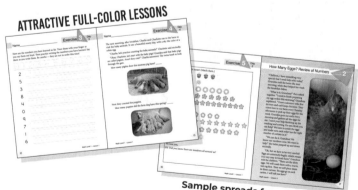

Sample spreads from Book 1

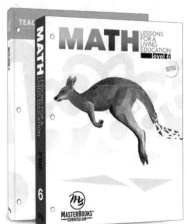

Level 6, Grade 6
978-1-68344-024-6

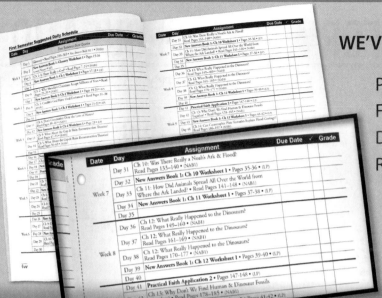